Problem Books in Mathematics

Series Editor
Peter Winkler, Department of Mathematics, Dartmouth College, Hanover, NH, USA

Books in this series are devoted exclusively to problems - challenging, difficult, but accessible problems. They are intended to help at all levels - in college, in graduate school, and in the profession. Arthur Engels "Problem-Solving Strategies" is good for elementary students and Richard Guys "Unsolved Problems in Number Theory" is the classical advanced prototype. The series also features a number of successful titles that prepare students for problem-solving competitions.

Bálint Hujter • Dániel Lenger • Gábor Szűcs

Mathematical Explorations for Young Minds

Adventures on the Island of Oxisz

Bálint Hujter
Budapest, Hungary

Dániel Lenger
Budapest, Hungary

Gábor Szűcs
Budapest, Hungary

ISSN 0941-3502   ISSN 2197-8506 (electronic)
Problem Books in Mathematics
ISBN 978-3-031-64184-8   ISBN 978-3-031-64185-5 (eBook)
https://doi.org/10.1007/978-3-031-64185-5

Mathematics Subject Classification: 97D50, 97M99, 97U40

Translated by Zoltán Gyenes. English translation of the original Hungarian edition published by Typotex, Budapest, 2022.

Translation from the Hungarian language edition: "Hajók, festmények, nagymamák: Matematikai kalandok Óxisz szigetén" by Bálint Hujter et al., © 2022. Published by Typotex. All Rights Reserved.

© The Editor(s) (if applicable) and The Author(s), under exclusive license to Springer Nature Switzerland AG 2024

This work is subject to copyright. All rights are solely and exclusively licensed by the Publisher, whether the whole or part of the material is concerned, specifically the rights of reprinting, reuse of illustrations, recitation, broadcasting, reproduction on microfilms or in any other physical way, and transmission or information storage and retrieval, electronic adaptation, computer software, or by similar or dissimilar methodology now known or hereafter developed.

The use of general descriptive names, registered names, trademarks, service marks, etc. in this publication does not imply, even in the absence of a specific statement, that such names are exempt from the relevant protective laws and regulations and therefore free for general use.

The publisher, the authors and the editors are safe to assume that the advice and information in this book are believed to be true and accurate at the date of publication. Neither the publisher nor the authors or the editors give a warranty, expressed or implied, with respect to the material contained herein or for any errors or omissions that may have been made. The publisher remains neutral with regard to jurisdictional claims in published maps and institutional affiliations.

This Springer imprint is published by the registered company Springer Nature Switzerland AG
The registered company address is: Gewerbestrasse 11, 6330 Cham, Switzerland

If disposing of this product, please recycle the paper.

Preface

The fabled island of Oxisz is not celebrated for its natural wonders or architectural grandeur. Rather, it is renowned for the mathematical puzzles with which the islanders entertain their visitors. We invite our readers to embark on 13+1 journeys during which we showcase 84 of our favorite Oxiszian puzzles. We trust our readers will enjoy thinking about these puzzles and discover solutions on their own or with the aid of the hints (see Part 2).

In the Solutions section, we explore the discussions among three friends—Albrecht, Tarkal, and Zsordi—while they hiked together on the island. We encourage our readers to delve into their adventures, but only after thoroughly contemplating the problems. After all, every experienced traveler knows: reading a travelogue cannot compare to the joy of personal discovery.

The majority of the 84 problems were originally introduced to students aged 12–18 at the Dürer Math Competition in Hungary. Many of these puzzles were invented by the competition organizers (including the three authors), while others were shared by our peers or sourced from various problem books. These puzzles have been used in math camps, problem-solving circles, and occasionally in school math classes as well. Hence, our experiences related to the puzzles are diverse. Some are memorable for the puzzle's inception, others for the ensuing discussions, or perhaps for the solutions proposed by our students. These moments are echoed in the adventures of Albrecht, Tarkal, and Zsordi.

We happily recommend the puzzles to anyone interested, as they do not require any specific mathematical background. Familiarity with the mathematics curriculum up to the first six grades is adequate for the initial chapters. The latter half of the book features progressively more challenging problems. However, these still do not demand additional schooling but rather depend on solutions to earlier problems, creative ideas, and persistent thinking.

The last problem of each chapter is a two-player strategic game. These are suitable for challenging a friend or relative to a match played according to the described rules.

We wish all our dear readers joyful hiking and contemplation!

Figure 1

Regarding how Albrecht, Tarkal, and Zsordi (Fig. 1) stumbled upon the island of Oxisz, the details remain unclear. What we do know for certain is that after competing together in the final stage of the Dürer Competition in Miskolc, they set their course northeastward...

Budapest, Hungary	Bálint Hujter
Budapest, Hungary	Dániel Lenger
Budapest, Hungary	Gábor Szűcs

Acknowledgments

The defining portion of this book consists of problems set originally in the first 15 seasons of the Dürer competition. The conception of these puzzles would be inconceivable without the devoted efforts of the competition's organizers and the stimulating atmosphere fostered among them. When two organizers meet, the following scene often unfolds:

"Hi! I came up with a Dürer problem. Are you interested?"

"Of course, I'd be happy to think about it. But then, I'll share one too!" Countless ideas and puzzle proposals emerge from collaborative brainstorming sessions, which occasionally extend well into the late hours of the night. These proposals are refined during problem committee meetings to ensure that participants can find much joy in them.

Therefore, we would like to express our heartfelt gratitude to all current and former organizers of the Dürer competition, who serve as our co-creators. Within the chapter titled "Sources" readers may encounter the names of numerous individuals. We would like to highlight Kartal Nagy, whose unique task ideas form a characterizing portion of the book.

For many years, Júlia Szűcs was responsible for the artistic design of the competition's certificates, posters, and other graphics. It brings us great joy that she also created illustrations for this book.

The original Hungarian version of this book was published in 2022 by Typotex Publishing House. We would like to express our gratitude to them, and extend our special thanks to Katalin Fried, the editor of the Hungarian version. Her comments and suggestions greatly influenced the manuscript, resulting in significant improvements. The structure became more cohesive, and the solutions more transparent as a result. Furthermore, based on her recommendations, the book has been enriched with new questions and solutions.

We extend our gratitude to our friends, all of whom are experienced organizers of the competition, namely Kristóf Huszár, András Imolay, Viktor Kiss, and Lilla Tóthmérész, for reviewing certain chapters of the manuscript.

The creating of this book has been a long-planned endeavor, made possible by the generous support of Sonrisa Technologies and The Joy of Thinking Foundation. We have enjoyed the complete trust of both Miklós Szurdi and Péter Juhász, the leaders of the two organizations, throughout the sometimes protracted, creativity-filled process marked by occasional revisions.

The translations to English were completed by our friend Zoltán Gyenes. We would like to express our heartfelt gratitude for his dedicated effort.

This book was funded by the Content Pedagogy Research Program and the Education Research Program of the Hungarian Academy of Sciences.

About the Authors

Bálint Hujter is a mathematics teacher for special math classes at Fazekas Mihály Gimnázium, Budapest.

Dániel Lenger is a mathematician and a math teacher. He teaches at Eötvös Loránd University, Budapest, and Fazekas Mihály Gimnázium, Budapest.

Gábor Szűcs is a mathematics teacher at Joy of Thinking Foundation and an assistant research fellow at Rényi Alfréd Mathematical Institute, Budapest.

The authors began organizing the Dürer competitions while studying mathematics at Eötvös Loránd University. Gábor was one of the founders of the competition in 2007, while Bálint and Dániel also served as leaders in the organizing team from 2008 to 2020. All three authors are also involved in other programs for mathematically gifted high school students, including Pósa camps, Math Olympiads, and KöMaL (Mathematical and Physical Journal for Secondary Schools). Illustrations of the book are made by **Júlia Szűcs**, who had been the chief visual designer of the Dürer Competition's material for many years. Translations to English were made by **Zoltán Gyenes** (math teacher at Fazekas Mihály Gimnázium).

Contents

Preface	v
Acknowledgments	vii
About the Authors	ix
Problems	2
1. Discover the Island of Oxisz	2
2. Painting with a Boat and a Barrel	5
3. Skiing Between Fréz and Gard	8
4. Football in Brassó	11
5. Can I Lock the Cat in the Mill?	14
6. The Trick to Making Moles Disappear	17
7. The Rejuvenating Grandma Is Baking a Pie	21
8. Reconcilable Differences	24
9. How Many Centimeters Is a Dessert?	27
10. Self-intersecting Tic-Tac-Toe	29
11. Flowers on the Tiles of the Housing Estates	31
12. Trolleybus on the Oktogon	34
13. Bloody Serious Problems	37
14. Icing on the Cake—Really Difficult Problems	40
Hints	42
15. Hints to the Problems	43

Solutions 50

1. Discover the Island of Oxisz **51**

2. Painting with a Boat and a Barrel **60**

3. Skiing Between Fréz and Gard **67**

4. Football in Brassó **76**

5. Can I Lock the Cat in the Mill? **87**

6. The Trick to Making Moles Disappear **94**

7. The Rejuvenating Grandma Is Baking a Pie **106**

8. Reconcilable Differences **117**

9. How Many Centimeters Is a Dessert? **132**

10. Self-intersecting Tic-Tac-Toe **142**

11. Flowers on the Tiles of the Housing Estates **154**

12. Trolleybus on the Oktogon **165**

13. Bloody Serious Problems **173**

14. Icing on the Cake—Really Difficult Problems **179**

Sources **195**

Problems

1. Discover the Island of Oxisz

1.1 Embassies on the Island of Oxisz

The map displays the 14 countries situated on the Island of Oxisz (Fig. 2).

Bergengocia, a country located on a different island, aims to establish embassies on the Island of Oxisz. Those countries are considered partners of Bergengocia that have an embassy of Bergengocia or share border with a country that has an embassy of Bergengocia.

Find the smallest number of countries with an embassy of Bergengocia such that every country on the Island of Oxisz is a partner of Bergengocia.

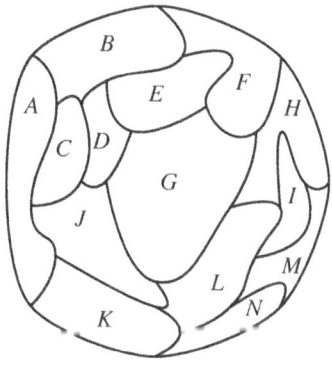

Figure 2

1.2 Zoo Runway

Rhinos, hippos, and flamingos have arrived at the Oxisz Zoo. Using an old football field they create a common space for them (Fig. 3). While the lengths of some sides are known, others are missing. Find the total length of the surrounding fence in meters.

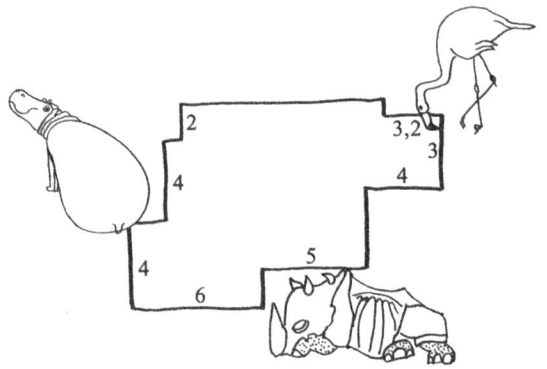

Figure 3

1.3 Incomplete Divisor Sequences

Porca was practicing division and enumerated all the divisors of a number in increasing order (for easy reference we will call these sequences *divisor sequences*). For example:

$$42: 1, 2, 3, 6, 7, 14, 21, 42.$$

Porca's little brother is not familiar with numbers, so he considered Porca's notes as something to scribble on. Unfortunately, the little boy also scribbled on a large part of the divisor sequence below:

$$\spadesuit: 1, \blacksquare, \blacksquare, 5, \blacksquare, \blacksquare, \blacksquare, \blacksquare, \blacksquare.$$

Find all the divisors that were scribbled on.

1.4 Office

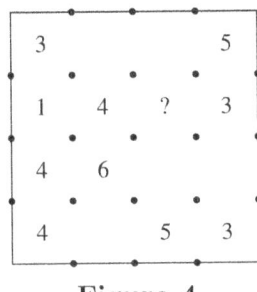

Figure 4

Figure 4 depicts the floor plan of an office divided into 16 square-shaped cubicles. Each cubicle houses a single person. The walls of the cubicles can be either transparent or opaque. Each worker tallies the number of people he can see in the same row and column. The image displays some of these values. How many people does Litor see who works in the cubicle indicated by the question mark?

1.5 Bishops and Rooks

Is it possible to place:

  a) Four bishops and four rooks   *b)* Five bishops and five rooks

on a chessboard such that none of the pieces can capture another piece?

Figure 5

We say that a piece can capture another if it can move with a single legal move to the square occupied by the other piece. For rooks, this includes horizontal or vertical moves; for bishops, it involves diagonal moves. The colors of the pieces are ignored, so, for example, a black piece can capture another black piece (Fig. 5).

1.6 Rock–Paper–Scissors with a Twist

On the Island of Oxisz, a different version of the game of Rock–Paper–Scissors became popular. In this version of the game each player has three cards: one with a picture of a rock, another with a picture of a piece of paper, and a third one with a picture of scissors. In the beginning of the game, both players place their three cards in front of them. Subsequently, the two players take turns picking a card from the other player until each is left with a single card. Finally the remaining cards are compared using the usual Rock–Paper–Scissors rule to determine the game's winner. If the two cards are the same, the game ends in a draw.

Determine whether this is a fair game.

2. Painting with a Boat and a Barrel

2.1 Lifeboat

A ship is sinking (Fig. 6). In the lifeboat there is room for five people, and it takes 3 minutes to reach the nearest island. What is the largest number of people that can be saved if the ship sinks completely in 20 minutes?

Figure 6

2.2 Barrels of Wine

In a wine cellar there are two identical barrels (Fig. 7). One of them is filled with wine, weighing a total of 188 kg, while the other one is half-filled with wine, weighing a total of 115 kg. Determine the weight of an empty barrel.

Figure 7

2.3 Train Cars

Trains on the Trans-Oxis line have to satisfy the following conditions:

- Between any two second class cars, there must be a first class car.
- Between any two first class cars there must be a sleeping car.
- Between any two sleeping cars there must be a dining car.
- The number of dining cars in a train is at most two.

What is the maximum number of cars in such a train, if we do not count the locomotive(s) and there are no other types of cars?

2.4 Contemporary Painter

A contemporary painter creates his main work titled *Big black rectangle* depicting a 5 × 9 black rectangle (Fig. 8).[1] His work becomes immensely successful leading to invitations to exhibitions and interviews. He also earns a lot of money: He eats caviar for lunch and truffles for dinner.

Figure 8

Unfortunately he does not produce another successful painting. He borrows money from friends, and at the time of our story, he owes money to nine people, some of whom want to see him in prison.

To resolve his debt he decides to divide his famous painting among them. However, he imposes two conditions: The pieces must be rectangles with integer sides, and no two rectangles can be identical (congruent):

a) Can we assist the painter in finding a division that meets the conditions?

b) Is it possible for the painter to retain a piece for himself? In other words, is it possible to divide a 5 × 9 rectangle into ten distinct rectangles?

---

[1] The story of the problem was inspired by Kazimir Malevich's painting titled "Black Square."

2.5 The Logo of the "Super(stitious)man"

Figure 9

The painter we met in the previous problem has been asked to design a logo for a new toy called "Super(stitious)man."

To design the logo, he is using equilateral triangles and squares, each with a side length of 1 cm, to create 2D shapes with a combined perimeter of 13 cm. The triangles and squares in the 2D shape are glued together along their entire sides without any overlap (Fig. 9):

Can such a shape be constructed by:

a) Using a single triangle and some squares?

b) Using the same number of squares and triangles?

c) Using only triangles?

d) Using only squares?

2.6 The Game of Covering Numbers

A table contains numbers ranging from 1 to 10 (see Fig. 10). Two players take turns covering a number until only two numbers remain. If the sum of the remaining two numbers is even, the first player wins, and if it is odd, the second player wins.

How would you play this game if you also get to choose whether to be the first or the second player?

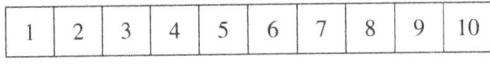

Figure 10

3. Skiing Between Fréz and Gard

3.1 Skiing Competition

In a skiing competition, participants take turns sliding down the slope, with the next contestant starting once the previous one reaches the finish line. In the live broadcast, whenever a contestant completes their run, their name and a number indicating their current position are displayed (Fig. 11). In the competition featuring eight contestants, the following names and numbers were showcased during the broadcast:

Reichelt 1; Svindal 1; Jansrud 2; Pinturault 3;
Küng 4; Hirscher 1; Ligety 3; Miller 2

Which contestants took the podium at the end of the competition?

Figure 11

3.2 Shopping Carts

In the parking lot of the Common Sense Supermarket the shopping carts are kept pushed together. If seven shopping carts are pushed together, their total length is 3 meters. If 10 shopping carts are pushed together, their total length is 4 meters. Determine the length of a single shopping cart. (*Fig. 12 is not to scale.*)

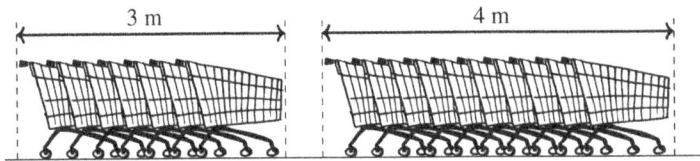

Figure 12

3.3 Connecting a Grid into a Polygon

Connect the points to get a polygon. All the vertices of the polygon must be chosen from the 20 points below. Maximize the number of the sides.

For example, the polygon on the right side of Fig. 13 has 11 sides.

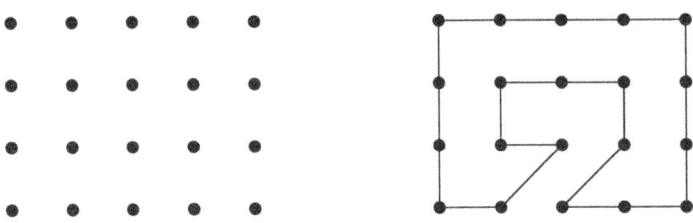

Figure 13

3.4 Buses Connecting Villages

On the map below there are nine villages (with names Abél, Begó, Cilk, Dénő, Enci, Fréz, Gard, Hóly, Ibol) and some roads connecting them. Unfortunately, numbers were used on the map instead of the names of the villages (Fig. 14).

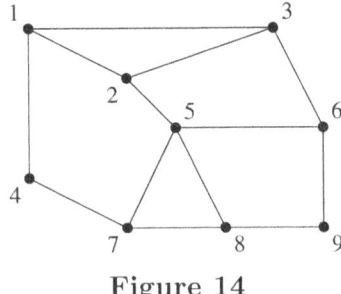

Figure 14

We know that there are three circle routes between some of the villages:

- Bus no. 41 through villages Abél–Begó–Cilk–Dénő–Enci–Abél
- Bus no. 42 through villages Abél–Fréz–Gard–Begó–Dénő–Enci–Abél
- Bus no. 43 through villages Begó–Gard–Hóly–Ibol–Cilk–Begó

Buses only travel on the roads depicted on the map and will not pass through a village without stopping. Find which number corresponds to which village.

3.5 Who Is My Bigger Neighbor?

a) Nine children sit around a circular table, and each gets one of the nine cards containing the integers from 1 to 9. They glance at the cards held by their two neighbors and point to the child holding the higher number (Fig. 15). Is it possible to arrange the nine cards so that only one child points to the left, and everybody else points to the right?

b) What would be the answer to the previous question, if eight children would sit around the table, and the card with number 9 will not be used?

Figure 15

3.6 Placing Tokens in Four Squares

There is a board with four squares, and there are two players, each with three tokens (Fig. 16).

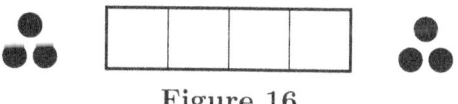

Figure 16

The two players take turns to place one of their tokens in one of the squares on the board. The second player wins the game if all four squares contain a different number of tokens.

How would you play this game if you also get to choose whether to be the first or the second player?

4. Football in Brassó

4.1 Stagecoaches

"Well, I want you to answer me this. If two stage-coaches leave Pozsony for Brassó every day, and as many leave Brassó for Pozsony, and assuming that the journey takes ten days, then how many coaches would you meet on the way in one stage-coach from Pozsony to Brassó, if you travelled in one stage-coach from Pozsony to Brassó?"

4.2 Pyramid of Numbers

Figure 17 depicts a pyramid of numbers. Fill in the empty cells with positive integers such that every number (excluding the last row) is the sum of the two numbers below it.

How many different solutions are there?

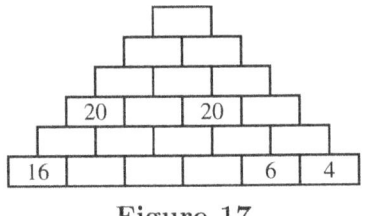

Figure 17

4.3 Photocopied Polygon

I drew the following 9-gon in a square lattice (Fig. 18).

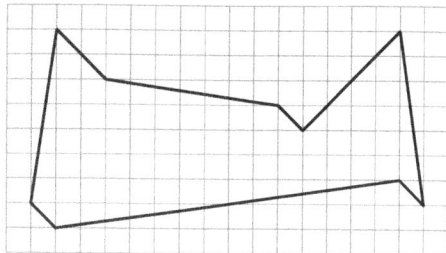

Figure 18

I made a photocopy of my drawing, but the machine zoomed in slightly (keeping the shape of the 9-gon), such that the length of the shortest sides became exactly 1 cm, and the length of the longest side became exactly 10 cm. Find the perimeter of the 9-gon in the photocopied picture.

4.4 Polygon with Many Concave Angles

What is the biggest possible number of concave angles in a 12-gon?
Angles greater than 180° are called concave angles.

4.5 Mixed up Kits

The football team of Oxisz has 11 players, each with a jersey and shorts numbered from 1 to 11 (Fig. 19).

Figure 19

However, before today's practice the shorts got mixed up, causing players to wear shorts with a different number from their jerseys:
 a) For each player we add up the numbers of their shorts and jerseys. Is it possible that we get 11 consecutive numbers?
 b) The next day the shorts were mixed up again, but the goalkeeper was careful enough to take the shorts and the jersey labeled with 1, and only the remaining players had their shorts mixed up. We add up the numbers on the shorts and jerseys of the 10 outfield players. Is it possible that we get 10 consecutive numbers?

4.6 Placing Tokens on a Board with Five Squares

In this game 10 tokens will be arranged on a board containing five squares. Before starting 1 of the 10 tokens is placed in the middle square on the board, and then three tokens are given to the first and six tokens are given to the second player (Fig. 20).

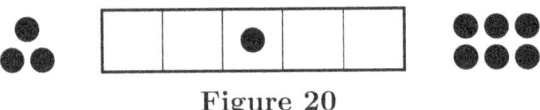

Figure 20

The game consists of three rounds. In each round the first player places one token in one of the squares on the board, followed by the second player placing two tokens. These two tokens can either be placed in the same square or in two different squares. At the end of third round all the tokens are on the board, and the winner is decided. If every square has a different number of tokens, the second player wins. However, if two squares can be found with an equal number of tokens, the first player wins.

How would you play this game if you also get to choose whether to be the first or the second player?

5. Can I Lock the Cat in the Mill?

5.1 Magic?

Do the following operations in your notebook. Think about the day of the month you were born. Add 17 to this number, and multiply the result by 2. Subtract the sum of the digits from the result. Next, add together the digits of the resulting number, multiply the sum by 4, and finally, add 6 to the result. Did you get 42? How did we guess it?

5.2 Tortoise, Cat, Table

If I place my cat on my table, its head is 115 cm above the head of my tortoise on the ground. If I place the tortoise on my table and the cat on the ground, the head of my tortoise is 65 cm above the head of the cat (Fig. 21). Find the height of my table.

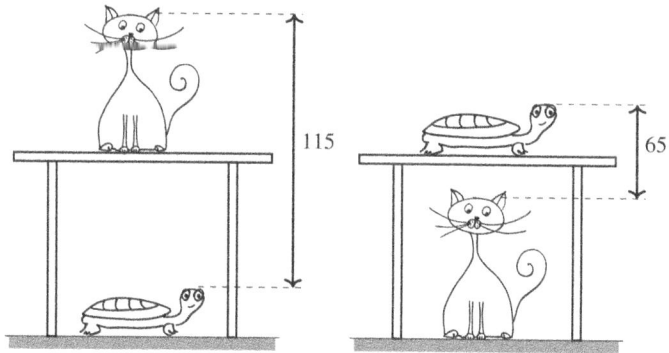

Figure 21

5.3 Central Lock

My car is equipped with a central locking system that operates in three states: In state (C) the lock is closed, in state (O) the lock is open, in state (T) only the trunk can be opened, while the doors remain closed. The lock is operated with a remote featuring a single button. By pressing the button repeatedly, the system cycles between states (O), (C), and (T). Additionally, if the lock is in state (O), and the button is not pressed for a whole minute, the lock automatically switches to state (T) (Fig. 22).

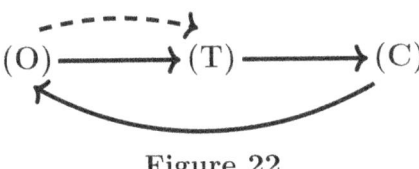

Figure 22

Unfortunately I cannot recall if I locked the car when I left it in front of the house. I do not want to go back to the car to check, but luckily the remote works even from the window. What is the minimum number of button presses required to surely put the lock into state (C)?

Initially, my car can be in any of the three states, and distinguishing between the states is not possible by simply observing the car from the window.

5.4 Nine Men's Morris Without a Mill

Nine Men's Morris is a two-player game played on a board, which is depicted in Fig. 23. This time we will play the game solo, with no opponent. The objective is to place pieces of the same color on the board in a way that no three pieces align to form a mill (i.e., three pieces along the same line segment). Find the maximum number of pieces that can be positioned on the board following these rules.

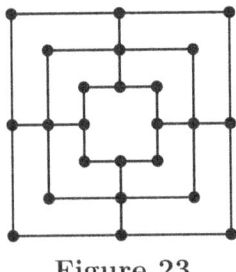

Figure 23

5.5 Length of Line Segments in a Circle

Find the total length of the line segments drawn in thick in Fig. 24, if the radius of the circle is 4 cm.

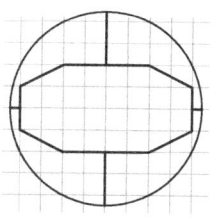

Figure 24

5.6 Jump over the Other Piece!

A white piece is placed on one of the first three squares of the playing field that can be seen in Fig. 25. Similarly, a black piece is being placed on one of the last three squares. The first player controls the white piece, while the second player controls the black piece. Players alternate turns to take one or two steps toward the other piece. Stepping on the other player's piece is not permitted. The winner is the player who manages to jump over the piece of the other player.

How would you play this game if you can also decide whether to be the first or the second player?

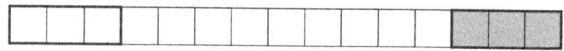

Figure 25

6. The Trick to Making Moles Disappear

6.1 Blocks of Light Bulbs

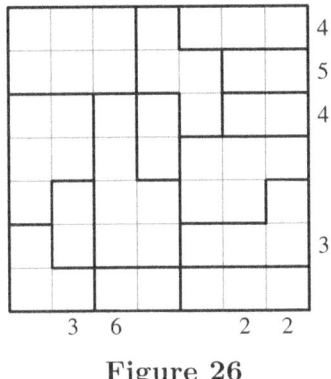

Figure 26

In Fig. 26, there is a light bulb in each of the 49 small squares. The light bulbs are divided into blocks by the thick lines. In a block either all or none of the light bulbs is lit. The numbers featured at some of the rows and columns show the total number of lit light bulbs in them.

How many light bulbs are lit on the board?

6.2 Rectangle with a Zigzag

We divide a rectangle-shaped paper along a zigzagging line into triangles, see Fig. 27. The numbers written in five of the triangles denote their respective areas, measured in a certain unit of area. Find the area of the sixth triangle.

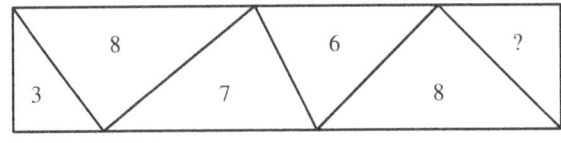

Figure 27

6.3 Group Stage Qualification

In a soccer tournament there are four teams in a group. Within a group, every team plays against every other team exactly once. Teams earn 3 points for a win, 0 points for a loss, and 1 point for a draw. The two teams with the highest scores progress to the next round, while the remaining two are eliminated (ties are resolved by drawing lots). Before the start of the tournament, the coach of the team writes the following statements on the board:

1. We need at least A points to progress to the next ground.

2. We need at least B points to have a chance to progress to the next ground.

3. The highest score with which it is still possible to be eliminated: C.

4. The highest score with which we are guaranteed to be eliminated: D.

Find the values of A, B, C, and D supposing that the coach made no mistakes.

6.4 Mole and Bulldozer

Farmer Brown's garden is divided into a neat 4×4 grid. Unfortunately, a pesky mole has left mole-hills in some squares. To tackle the problem, Farmer Brown hops onto his bulldozer and drives it along a column from North to South and then along another column in the opposite direction (i.e., from South to North). Afterward he drives along a row from East to West, and finally he drives along another row in the opposite direction (i.e., from West to East).

Find the smallest number of mole-hills that is impossible to be bulldozed by Farmer Brown in this manner.

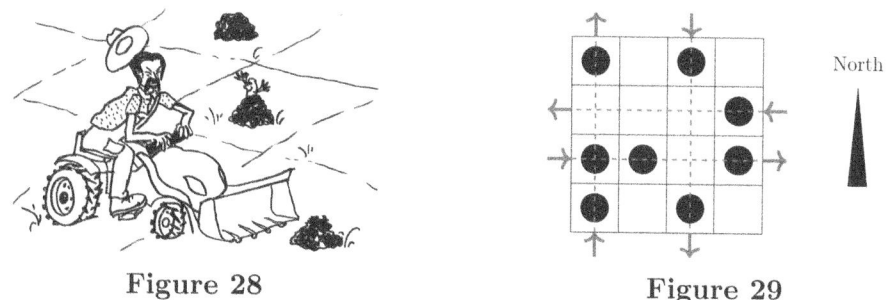

Figure 28 Figure 29

Fig. 28 depicts Farmer Brown on his bulldozer, while Fig. 29 shows eight mole-hills that Farmer Brown can bulldoze by following the arrows.

6.5 The Magic Trick of Rodolfo

We would like to understand an illusion of Rodolfo.[2] In this illusion, each member of the audience is seated at their own table with a hat and a stack of four cards: the Aces of Hearts, Bells, Leaves, and Acorns.[3]

Initially, the members of the audience select some of their cards—it could be all, none, or just a few—and toss them into their hat, forming a deck with the remaining cards on their table. The illusion unfold over multiple rounds. Rodolfo picks an Ace from his own cards and places it prominently, where it is easily seen. In this round the card chosen by Rodolfo becomes the magic card.

The audience has to perform one of the following acts:

- If the magic card is in the hat, it has to be placed on the top of the pile on the table.

- If the magic card is on the top of pile, it has to be placed in the hat.

- If the magic card is in the pile, but not on the top, then it has to be placed on the top of the pile.

At the end of the performance, all members of the audience are surprised to discover that every single one of their cards has ended up inside the hat.

What can be the secret behind this illusion? Find a sequence of the suits that enables the illusion to be performed. Try to keep your sequence of suits as short as possible.

6.6 The Game of Turning Tokens Upside Down

At the beginning of the game, we have some tokens with a blue and a red side. Some tokens have their red side up, while others display their blue side. The player taking their turn can choose from the following possibilities:

- He can take away 1 or 2 blue tokens from the table.

- He can turn over 1 or 2 red tokens, thus making them blue.

The winner is the player taking away the last token. How would you play this game?

---

[2] Rodolfo (Rezső Gács, 1911–1987) remains one of the most renowned Hungarian magicians. Please note that the illusion described in the problem is purely imaginary and has never been performed in real life.

[3] These suits are from the German-suited playing cards, popular in many parts of Central Europe.

You can decide whether you want to be the first or the second player after checking the number of red and blue tokens, respectively (Fig. 30).

Figure 30

7. The Rejuvenating Grandma Is Baking a Pie

7.1 Grandma Is Getting Younger

My grandma thinks she is getting younger, because 5 years ago she was five times as old as me, and now she is only four times as old as me:

a) How old is my grandma?

b) In how many years will she be three times as old as me?

7.2 Too Many Knaves

The ship *Graceful Hippopotamus* is manned by a crew of 21 (including the captain), all of whom are either knights or knaves.[4]

After the ship reached the Island of Oxisz after three months at the sea, the crew disembarked one by one, the captain being the last to leave the ship.

Upon leaving the ship, all members of the crew except for the captain stated the following: *"More knights than knaves remained on the ship."*

How many knaves are on duty on the *Graceful Hippopotamus*?

7.3 Finding the Missing Sums

Four different integer numbers are given, and we calculate all six possible pairwise sums of them. We get the following values in increasing order: 105, 111, 112, ■, ■, 120. Find the two missing sums.

---

[4] Knights are always telling the truth, while knaves are always lying. They were introduced in Raymond Smullyan's book *What Is the Name of This Book* and have since become part of the mathematical folklore, perhaps partly due to their catchy alliterating names.

7.4 Dividing the Pie

Granny's pie is a square measuring 30 cm × 30 cm. The most delicious part of the pie is its crust. Three grandchildren wish to split the pie evenly so that each of them gets an equal share of both the pie and its crust (Fig. 31.) Is it possible to divide the pie equally in this manner?

Figure 31

7.5 Crouching and Standing Students

Ten students with varying heights play the following game in the gym. Initially, they are all crouching. Their PE teacher claps once per second. After each clap, the smallest crouching student stands up, while any standing students shorter than him return to the crouching position (Fig. 32). The game continues until all students are standing:

a) After how many claps of the PE teacher will the game conclude?

b) How many students will be standing after the 200th clap?

Figure 32

7.6 The Game of Two Heaps

There are two heaps of tokens on the table. Two players take turns following these rules: The player whose turn it is removes one of the two heaps from the game and divides the other heap into two smaller heaps (see Fig. 33). Dividing a heap is only allowed if it contains at least two tokens. After each move, two heaps must remain, each containing at least one token. The player unable to make a legal move loses the game.

How would you play this game if you can decide whether to be the first or the second player based on the initial position?

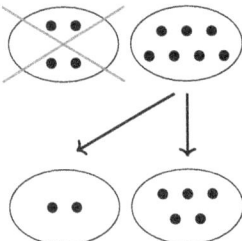

Figure 33

8. Reconcilable Differences

8.1 Erring Organizers?

During a meeting of the organizers (Fig. 34) of the Dürer competition the following conversation took place:

Zsófi: Dürer is the best competition in the world.
Bálint: Zsófi is wrong.
Gábor: Zsófi is wrong.
Kartal: Bálint is wrong.
Dani: Gábor is wrong.
Benedek: Gábor is wrong.
Magdi: Bálint is wrong.
Juli: Kartal is wrong.
Bea: Benedek is wrong.
Peter: Dani is wrong.

How many of the organizers are wrong?

Figure 34

8.2 Find the Connection

a) How many seven-digit numbers can be created using five digit 5s and two digit 2s?

b) How many ways can one travel from point A to point B in Fig. 35, while following the lines and only moving rightward or downward?

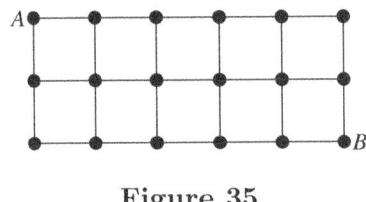

Figure 35

c) Find the maximum number of points of intersection between seven lines.

8.3 Without Adjacent Obtuse Angles

Find the maximum number of vertices in a convex polygon that has no two adjacent obtuse angles.

8.4 Quadrilaterals That Are Small and Big at the Same Time

a) Can a quadrilateral have all sides longer than 10 cm, but an area smaller than 1 cm^2?

b) Can a quadrilateral have all sides shorter than 1 cm, but an area larger than 1 cm^2?

8.5 The Number of Odd Divisors

How many odd divisors can a number have that has 6 even divisors?

8.6 Policemen in Cubetown

One of the two players moves a thief, and the other player moves two policemen along the streets of Cubetown (see Fig. 36).

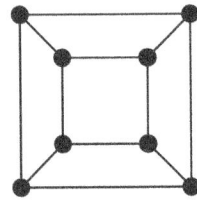

Figure 36

At the start of the game, each of the three characters is at a corner (marked with a dot). In each round of the game the thief makes a move to a neighboring corner, followed by the two policemen making a move in a similar manner. Once during the game, the thief is allowed to make a double move. The policemen win if there is a moment in the game when the thief and one of the policemen are at the same crossing. The thief wins if this does not happen in the first five rounds.

How would you play this game for the initial positions depicted in

a) Fig. 37,   b) Fig. 38,   c) Fig. 39,

if you can decide whether you want to play with the policemen or with the thief?

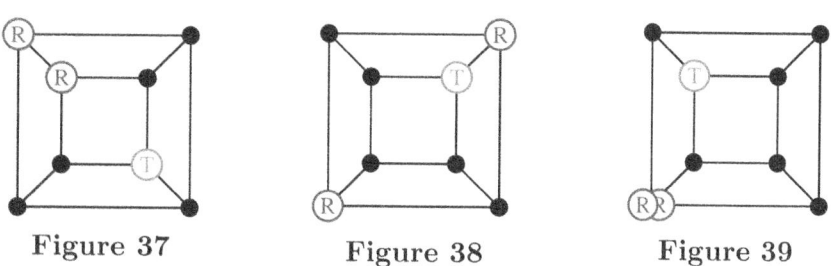

Figure 37   **Figure 38**   **Figure 39**

9. How Many Centimeters Is a Dessert?

9.1 Boys, Girls, T-Shirts

Ten children are sitting around a round table, comprising both boys and girls. Every boy's left neighbor wears a blue T-shirt, while every girl's right neighbor wears a green T-shirt. How many girls are sitting around the table?

9.2 Diagonal Halving an Octagon

A diagonal in a polygon is called "main diagonal," if it connects two opposite vertices. The main diagonals of a convex octagon all contained inside the octagon. What can be said about concave octagons? Is it true that every octagon has a main diagonal that is fully inside the octagon?

9.3 Missing Marks on a Ruler

A ruler currently has marks only at 0 and 12 cm. At least how many additional marks are needed on the ruler to make it possible to measure every integer distance up to 12 cm?

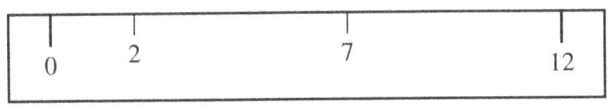

Figure 40

For example, if we add two marks at distances 2 and 7 cm from mark 0 (as in Fig. 40), we will be able to measure these five distances: 2 cm (as the distance between marks 0 and 2), 5 cm (as the distance between marks 2 and the 7, or marks 7 and 12), 7 cm (as the distance between marks 0 and 7), 10 cm (as the distance between marks 2 and 12), and 12 cm-t (as the distance between marks 0 and 12).

9.4 The Diet of Arthur Noodle

Arthur Noodle[5] is on a diet. After lunch he opts only for two types of desserts instead of five. He carefully considers which pair to choose, so he counts the number of calories contained in each pair. He obtains the following values in increasing order:

$$280, 300, 350, 400, 420, 450, 470, 520, 540, 590.$$

Determine the individual number of calories in each dessert.

9.5 Dissection into Rectangle

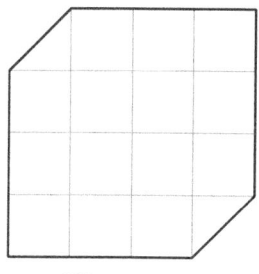

Figure 41

The shape shown in Fig. 41 was obtained by removing the opposite corners of a 4×4 square. Is it possible to cut the shape along a polyline into two pieces that can be rearranged to form a 3×5 rectangle?

9.6 Double Opening Move Tic-Tac-Toe

In a "double opening move" tic-tac-toe game on a 3×3 grid, the first player places two X's, followed by the second player placing an O. Subsequently, each player alternates turns until the playing field is completely filled. The first player wins if there are three X's that form a row, a column, or a diagonal, with no three O's forming a row, a column, or a diagonal. Otherwise the second player wins.

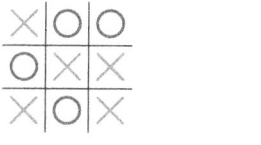 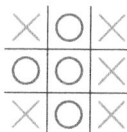 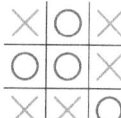

Figure 42   **Figure 43**   **Figure 44**

For example, in Fig. 42, the first player (X's) has won. In Figs. 43 and 44, the second player (O's) is the winner.

Which player has a winning strategy in this game?

---

[5]Arthur Noodle is a well known character in a Hungarian animated series for kids. He is famous for loving chocolate.

10. Self-intersecting Tic-Tac-Toe

10.1 Ice Creams on a Stick

At a shop, there is a special offer: The sticks of two ice cream bars can be exchanged for a new ice cream bar. We aim to make the most of this offer. Find the maximum number of ice cream bars we can get if we have:

  a) 20; b) 107; c) n sticks initially?

10.2 Enthusiasm and Half-heartedness

a) On the island of El Samolo, every resident is either half-hearted or enthusiastic. Once a visitor from a distant land arriving on the island was invited for dinner by a group of 10 residents. After dinner, the visitor asked all 10 members of the group about the number of enthusiastic inhabitants within their group.

He received the following answers: 3, 4, 5, 6, 7, 8, 9, 10, 11, 12.

Knowing that the answers of the half-hearted individuals cannot be more than the actual answer, and the answers of the enthusiastic individuals cannot be less than the actual number, determine the number of enthusiastic inhabitants within the group:

b) Can you find the answer in general? Is it possible to determine the number of enthusiastic residents in a group of 42 people, provided their answers are
$$a_1 \leq a_2 \leq \ldots \leq a_{42}?$$

10.3 Quadraples of Kritor and Lidek

Kritor has secretly chosen four integer numbers and noted down their pairwise sums on a piece of paper. Lidek has also chosen four integer numbers secretly and noted down their pairwise sums on another piece of paper. Is it possible that the six results on both pieces of paper are the same, while all the originally chosen eight numbers are different?

10.4 Uniquely Self-intersecting Polygon

A self-intersecting polygon is called *uniquely self-intersecting*, if each of its sides intersects exactly one other side, and none of its sides contains a vertex except for its endpoints:

a) Draw a uniquely self-intersecting polygon.

b) Determine the minimum number of sides of a uniquely self-intersecting polygon.

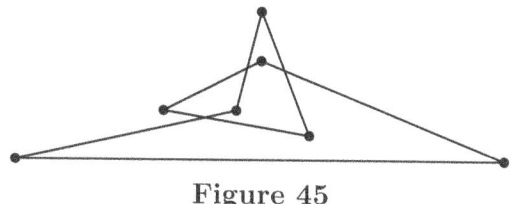

Figure 45

In Fig. 45 we can see a self-intersecting heptagon, which is unfortunately not uniquely self-intersecting, since its bottom side does not intersect any other of its sides.

10.5 Prime Orders of Divisors

If the divisors of a positive integer can be ordered in a way that the quotient of any two consecutive divisors is a prime, we call this order a *prime order*. Is it true that the divisors of every positive integer have a prime order?

10.6 Anti-Tic-Tac-Toe

In the game of anti-tic-tac-toe the first player places X's, while the second player places O's on a 3×3 board. The two players alternate turns, and the one who forms a row, a column, or a diagonal with his marks **loses** the game. If all nine squares are occupied and no such triple was created, the first player wins.

How would you play this game if you could decide whether to be the first or the second player?

11. Flowers on the Tiles of the Housing Estates

11.1 Housing Estate

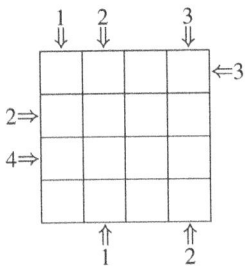

Figure 46

In a housing estate, 16 new houses are under construction in a 4×4 square grid shown in Fig. 46. Each house was designed with 1, 2, 3, or 4 floors. The architects paid extra attention to make sure that no two houses with the same height are in the same row or column. Kritor's family wants to buy a house, so they walked around the estate. Along the way, in certain rows and columns, they noted down the number of visible houses. A house is visible if no taller house obstructs its view.

Find the height of each house in the estate.

11.2 Bouquets of Flowers

At a flower shop, they sell three types of flowers: roses, tulips, and gerberas:

   a) How many bouquets of three flowers are there?

   b) How many bouquets of five flowers are there?

We consider two bouquets identical, if they contain the same number of flowers from all three types. A bouquet does not have to include all three types of flowers.

11.3 Tile

The congruent tiles in Figs. 47 and 48 contain quarter circles with their centers being the points marked in the picture. Determine the area of a tile, measured in cm², given that the height of a standing tile is 12 cm.

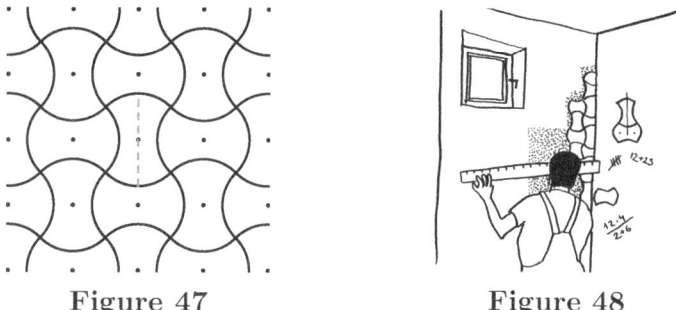

Figure 47　　　　　Figure 48

11.4 The Colorful Flag

The flag of a pirate ship is square shaped and consists of 4×4 small colored squares. Captain Mondrian instructed the crew to assign new colors to the small squares ensuring that each 2×2 part contains at least two squares of the same color. The crew (see Fig. 49) aims to create the most colorful flag possible. Find the maximum number of different colors that can be used on the flag.

Figure 49

11.5 Short Heights, Large Area

Is it possible to find a triangle with an area larger than 1 m², while all its heights are less than 1 cm?

11.6 Taking Away Tokens on the Same Line

Two players alternate turns to remove tokens from a rectangular arrangement on a table. On his turn, a player can remove a continuous line of tokens from a row or a column. The player taking away the last token wins the game.

How would you play this game for the following initial arrangements:

a) 16 tokens in a 4×4 lattice

b) 15 tokens in a 3×5 lattice

c) 20 tokens in a 4×5 lattice?

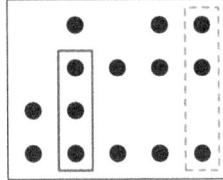

Figure 50

Figure 50 shows a position that was reached from the initial arrangement c). The player on turn can take away the three tokens highlighted with the continuous line. However, he cannot remove the three tokens highlighted with the dotted line, since there is gap between them.

12. Trolleybus on the Oktogon

12.1 Trolleybus, Bus, Tram

The following statements are true for the local public transport system of Oxisz:

   a) There is a city with trams but no trolleybuses.

   b) If a city has both buses and trams, it also has trolleybuses.

   c) At least one of the following two statements are true:

   i) No city lacks both buses and trolleybuses.

   ii) No city has both buses and trolleybuses.

   Is there a city having both trams and buses?

12.2 The Octagon Shaped Rug

Lidek's favorite rug has the shape of a regular octagon, and it is made of materials with three different colors (see Fig. 51). The borders separating the parts with different colors form a smaller and a bigger square. If we reflect the vertices of the inner white square across the sides of the larger square, we obtain four vertices of the octagon. Find the total area of the dark gray parts on the sides, given that the area of the light gray part is 1140 cm².

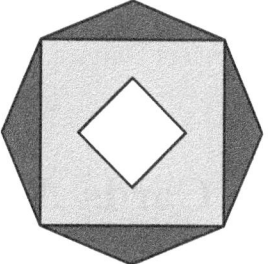

Figure 51

12.3 Filling in Tables with 1s and 2s

a) Is it possible to fill in a 2×2 table with digits 1 and 2 such that the four two-digit numbers we obtain by combining the digits in the rows and columns from left to right and top to bottom all have different remainders modulo four?

b) Is it possible to fill in a 3×3 table with digits 1 and 2 such that the six three-digit numbers we obtain by combining the digits in each row and column from left to right and top to bottom all have different remainders modulo six?

c) Is it possible to fill in a 4×4 table with digits 1 and 2 such the eight four-digit numbers we obtain by combining the digits in each row and column from left to right and top to bottom all have different remainder modulo eight?

12.4 Pairwise Products

a) We have computed the pairwise products of four positive numbers. Five of the six products are $20, 30, 40, 50, 60$. Find the value of the sixth product.

b) Also determine the original numbers.

12.5 Numbers with an Alternating Sequence of Divisors

A positive integer greater than 1 has an *alternating sequence of divisors* if enumerating the positive divisors in ascending order, the even and odd divisors appear alternately:

a) Show examples of numbers with an alternating sequence of divisors. Can you find an example with more than 10 divisors?

b) Is it possible to find a perfect square with an alternating sequence of divisors?

12.6 Duel of Five Cards Against Five Cards

Two players play the following game: Five cards are placed in front of each player, containing the numbers from 1 to 5. Then they take turns picking a card from their opponent, until a single card remains in front of both players. The game concludes at this point.

If the sum of the two remaining numbers is odd, the winner is the player with the larger number. If the sum of the two remaining numbers is even, then the winner is the player with the smaller number. If the two remaining numbers are equal, the winner is the first player.

How would you play this game if you also get to choose whether to be the first or the second player?

13. Bloody Serious Problems

13.1 Bidding on Each Other

A discussion is taking place between Dobda and Záhý.

- Záhý: In the last year's Dürer competition I scored three times as many points as you did.
- Dobda: Still, I scored more points than the number of counties in Hungary you know the capital of.
- Záhý: Fine, but I still know more than three times as many capitals as you do.
- Dobda: This is also true, but I know more capitals than the number of goals you scored in our last match.
- Záhý: Fine, but I scored more than four times as many goals in our last match as you did.
- Dobda: I admit that, however, I also scored in our last match.

Find the number of capitals Záhý knows, if it is also given that the maximum score in the last year's Drer competition was 60.

13.2 Queen, Rook, Bishop, Knight

Is it possible to place a queen, a rook, a bishop, and a knight in the upper-left quarter of the chessboard such that:

a) No two pieces can capture each other?

b) Every piece can capture exactly 1 other piece, and every piece can be captured by exactly 1 other piece?

c) Every piece can capture exactly 2 other pieces, and every piece can be captured by exactly 2 other pieces?

We say that a piece can capture another, if it can capture it with a legal chess move.

13.3 Polygon with Many Right Angles

Find the maximum number of right angles among a 12-gon's internal angles.

13.4 Lucky Numbers in Oxisz

In the horoscope of Oxisz each year is assigned with a lucky number. The lucky number of a given year is the largest odd divisor of the year (e.g., the lucky number of 2020 is 505).

What was the sum of the lucky numbers from year 1001 to year 2000?

13.5 Blood Donors and Raspberry Juices

This problem is based on a conversation between Kritor and Lidek.

"I've donated blood and noticed an interesting connection," said Kritor. "If A donates blood to B, then B will have some blood from A, while the reverse is not true." "I will say that A is a *blood donor* of B."

"Thus, being a blood donor is not reciprocal," observed Lidek.

"Exactly. However, if A donates blood to B, and then B donates blood to C, then A is also a blood donor of C," continued Kritor. "We could say that being a blood donator is hereditary."

"This is very intriguing, and reminds me of a question," continued Lidek. "Let's assume that three people intend to donate blood to each other with the only restriction that they cannot donate blood to a person who already has some of their blood." "In other words, they cannot donate blood to their blood donors."

After the donations, each person receives as many glasses of raspberry juice as the number of people he is the blood donor of:

a) Determine the maximum number of glasses of raspberry juice that three people can get.

b) What will be the answer if the number of people is six instead of three?

13.6 37 Tokens in Three Heaps

There are 37 tokens on the table divided into three heaps. Two players take turns to play the following game (Fig. 52): The player whose turn it is completely removes a heap and divides one of the other two heaps into two smaller heaps (see Fig. 53). A heap can be divided only if it contains at least two tokens. Thus, three heaps will remain after each round, each with at least one token. The player who cannot make a legal move loses the game.

Figure 52

How would you play this game if you can decide based on the initial position whether to be the first or the second player?

Some suggestions for initial positions: $(10, 12, 15)$, $(12, 12, 13)$, $(11, 13, 13)$. *The numbers in each triple stand for the numbers of tokens in each heap.*

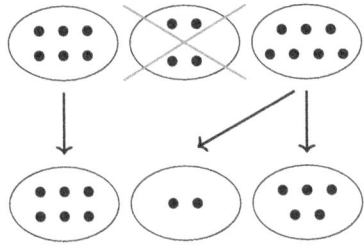

Figure 53

14. Icing on the Cake—Really Difficult Problems

14.1 Quadruples of Kritor, Lidek and Timol

Kritor, Lidek and Timol each thought about four numbers and noted the pairwise sums of their four respective numbers on separate pieces of paper. Given that their quadruples are all distinct, is it possible that all three pieces of papers contain the same results?

14.2 Meandering Zigzag

Let us consider polylines that do not intersect themselves and are not closed and call them *zigzags*. If a zigzag comprises at least 22 line segments, we call it *meandering*.

Find a meandering zigzag which can be covered with the smallest possible number of straight lines. How many lines are required?

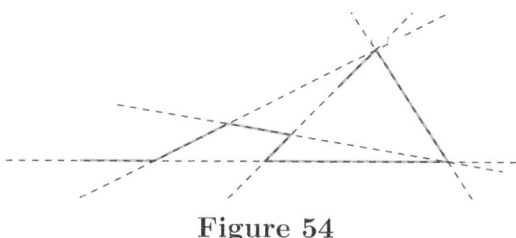

Figure 54

In Fig. 54, we can see a zigzag comprising seven line segments (therefore it is not yet meandering) which is covered with five straight lines.

14.3 Mole and Bulldozer

Farmer Brown has a square shaped garden that is divided into 12×12 smaller squares. A mole made some mole-hills on some of the squares, and therefore Farmer Brown bulldozes 6 rows and 6 columns.

What is the smallest number of mole-hills such that Farmer Brown cannot bulldoze all of them?

14.4 Dividing the Cake

Granny's pie is a rectangle measuring 18 cm × 36 cm. Its most delightful part is the crust, dipped in chocolate. Her three grandchildren wish to divide the pie equally so that each gets the same amount (area) of the pie, and each gets the same amount of the crust (perimeter):

- *a)* Is it possible to divide the cake into three convex pieces in such a manner?

- *b)* The whole family wants to eat from Granny's next pie; therefore they want to divide it into six convex parts, each containing the same amount of pie and also the same amount of crust.

- *c)* The pie gained fame throughout the entire neighborhood. Can it be divided among 12 people satisfying to the conditions above?

14.5 Last Digits of Divisors

Decide whether the following statement is true: Every integer has at least as many divisors ending in 1 or 9 as those in 3 or 7.

14.6 Heap Dividing Game, an Arbitrary Number of Tokens in 3 Heaps

A few tokens on a table are divided into three heaps. Two players take turns to play the following game: The player on turn removes a heap from the game and divides one of the other two heaps into two smaller heaps (see Fig. 55).

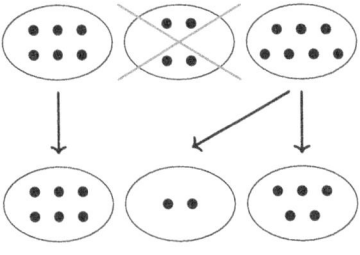

Figure 55

A heap can be divided only if it contains at least two tokens. Therefore, after each move three heaps remain, each with at least one token. The player who cannot make a legal move loses the game.

How would you play this game if you can decide based on the initial position whether to be the first or the second player?

Hints

15. Hints to the Problems

1.1. Even though country G in the middle of the island is big and has a lot of neighbors, putting an embassy here is not necessarily advantageous.

1.2. Complete the common space into a rectangle. What can be the perimeter of the rectangle?

1.3. Find the two divisors at the second and third positions. Try to take advantage of factor pairs. How is it possible that Porca found seven divisors?

1.4. Mark the walls that must be transparent or must be opaque. What conclusions can we draw about the row and column with the value of 6?

1.5. First, place the rooks such that they cannot capture each other. Now, where can the bishops be placed? It is also worth placing the bishops first and the rooks only afterward.

1.6. If the first player took the scissors from the second, then the second should take the paper from the first. Why?

2.1. The answer is 17. If your answer is different, try to find the flaw in your reasoning. There are several pitfalls where you can go wrong.

2.2. Why is the full barrel heavier, and by how much ? How much lighter would the barrels be if we were to remove half a barrel of wine from them?

2.3. Start with the last condition.

2.4 a) Such a division exists.
 b) No such division exists. When attempting to divide the painting into rectangles with small areas, what kind of issues may arise?

2.5. Before addressing the following questions, try to create shapes that meet the specified conditions. Try to find as many shapes as possible. Check whether they satisfy conditions *a) b) c)* or *d)*. If you have not found a solution for a question, there is a possibility that none exists. Investigate to discern the underlying reason.

2.6 Keep track of the number of odd and even numbers that are not yet covered after each move.

3.1. Find the positions of the contestants after Jansrud has reached the finish line. What are the positions after Pinturault's slide?

3.2. How does the length of a line of shopping carts change when three carts are added to the end?

3.3. Is it possible to include all the given points as a vertex of the polygon?

3.4. Sketch the routes of the buses through the villages. How many route segments branch out from village Begó?

3.5. The child with the smallest number will have their right neighbor pointing left, and the child with the biggest number will have their left neighbor pointing right.

3.6. Notice that the second player surely loses if there is a square with four tokens, or each square contains at least one token.

4.1. "...you would find the carriages that had left during the precious ten days..."

4.2. Begin by filling in the fields below the number 20 on the left.

4.3. Let us imagine that the figure is already zoomed. The length of the shortest side is 1 cm. To which other side can we compare it? We know that the longest side is 10 cm. The length of which other side can be determined from this?

4.4. The sum of the angles in a 12-gon is $10 \cdot 180°$. How many concave angles "fit into" this?

4.5. Find the sum of all the numbers on all the kits (shorts and jerseys) in both cases. Can it be the sum of 11 and 10 consecutive numbers, respectively?

4.6. Find a position at the end of the second round that guarantees a win for the second player in the end.

5.1. Do the calculations for different numbers! Compare the partial results!

5.2. Let us place the two pictures on the top of each other.

5.3. Answer a simpler question first: How can we ensure reaching state (T) by pressing the button twice?

5.4. How many pieces can be placed on a line segment? How many line segments are there?

5.5. Group the line segments in the diagram based on their directions.

5.6. Keep note of the number of empty squares between the two pieces after each move.

6.1. How can the third column contain exactly six light bulbs that are lit?

6.2. Create rectangles in the diagram.

6.3. Is it enough to win twice to surely progress to the next round? Is it possible to progress to the next round with no wins?

6.4. Consider the moment when the vertical (North and South) bulldozings have been completed, while the horizontal (East and West) bulldozings are yet to be done.
  How should the remaining mole-hills be arranged to obstruct Farmer Brown from bulldozing all of them?
  (Further hints can be found in the beginning of the solution.)

6.5. Find a sequence ensuring that a specific card emerges on the top of the pile.

6.6. How should we play if only blue tokens are left on the table? And if there are only red tokens?

7.1. Find the connection between our age difference and my age?

7.2. Let us work backward. What can we deduce when the captain is a knight and when a knave?

7.3. Can we determine the sum of the two smallest numbers? And the two largest?

7.4. It is not required that each grandchildren gets a single piece of the pie.

7.5. After how many claps will the second, third, fourth, and fifth smallest students stand up for the first time? And the highest?

7.6. The winning player will be the one who can create two heaps, each containing one token. Thus we cannot leave a heap with two tokens, because our opponent could divide it into two heaps with one token each. Try to decide whether we can leave a heap with 3 or 4 tokens.

8.1. What if Zsófi is right? What if Zsófi is wrong?

8.2. *a)* Arrange the numbers in increasing order. How many starts with digit 2? How many starts with digits 52? *b)* Calculate for each corner the number of ways to arrive there. *c)* Fix one of the lines. How many points of intersection can it have?

8.3. Find the maximum number of acute angles in a convex polygon.

8.4. *a)* Pick up four straws each longer than 10 cm. Arrange them to enclose the smallest possible area. *b)* Divide the quadrilateral into two triangles along a diagonal.

8.5. Try to find several examples. What can we observe about the number of odd divisors?

8.6. Color the corners using two colors following a "chessboard pattern."

9.1. The second neighbor to the right of a girl must be another girl. Why?

9.2. The statement is not true, and there exists a polygon with no main diagonal being fully inside the polygon.

9.3. How many new distances can become measurable by adding a single new mark?

9.4. We have encountered a similar situation in problem 7.3; therefore, it is worth using the ideas from it. Find the total number of calories contained in the five desserts. How can you take advantage of this to find the individual values?

9.5. Achieving our goal is possible. Let us align the target rectangle with the original hexagon. Let us start by cutting down smaller pieces, ensuring that the top right corners be perfectly aligned (see Fig. 56).

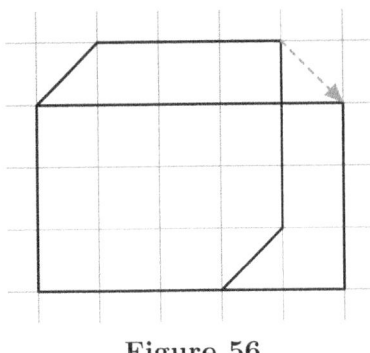

Figure 56

9.6. Although it's tempting to place the first X in the middle, maybe it's not a good idea.

10.1. Try out different methods. How many ice cream bars can you get?

10.2. The smallest answers definitely came from half-hearted members, while the largest ones were surely provided by the enthusiastic members. How can we separate the two groups from each other?

10.3. Try to create two groups of four from numbers $1, 2, \ldots, 8$ satisfying the condition.

10.4. Draw a self-intersecting polygon in which every side intersects *at least one* other side. Try to fix those sides that intersect multiple other sides.

10.5. It is worth experimenting with actual numbers. Start by ordering the divisors of $9, 16, 21, 24$ in the required manner. If successful, continue with 120 and 300.

10.6. Counterintuitively, it is advantageous for the first player to place his first X in the middle square.

11.1. What does it tell about the houses in a row if all four houses are visible from the street? What if only a single house is visible?

11.2. *a)* How many bouquets are there, where all three flowers are of the same type? How many bouquets are there, where all three flowers are of different types? Have we considered all cases?
   b) Let us count by analyzing the cases.

11.3. Divide the tile into smaller pieces and arrange them into a simpler shape with a more easily obtainable area.

11.4. Find a pattern with nine colors. 12 is an upper bound on the number of colors. Is it possible to use 10, 11, or even 12 colors?

11.5. Recall the solution of problem 8.4.

12.1. It can be proved that there has to be a city where trams are the sole mode of transportation. How does this helps us?

12.2. Divide the collection of the dark and light gray parts into congruent triangles.

12.3. Parts *a)* and *b)* cannot be solved; however, part *c)* can. For parts *a)* and *b)* it is worth checking where numbers with a single digit (i.e., 11, 22 and 111, 222) can appear.

12.4. Apply ideas from problem 7.3.

12.5. (*a*)) 6 has an alternating sequence of divisors: $1, 2, 3, 6$. How will the sequence of divisors change after multiplying it with a new prime factor? (*b*)) Is it possible to find a number divisible by 4 with an alternating sequence of divisors?

12.6. Try to find a connection to problem "Rock-paper-scissors with a twist" (1.6). If the card with 1 is associated with rock, what numbers can be associated with paper and scissors?

13.1. How does the answer to a question affect the answers to the others?

13.2. In parts (*a*)) and (*b*)) the pieces can be arranged on the board. In part (*c*)) it is worth considering the following questions: Who can capture the knight? Who can be captured by the knight?

13.3. Can it have twelve, eleven, ten right angles? And nine? Can you draw such an example?

13.4. We have written down the lucky numbers of all the years between 1001 and 2000. Can the same number occur multiple times?

13.5. Since there are three people, they cannot get more than $3 \times 2 = 6$ glasses of juice (which would be 30 glasses for six people). These bounds can be almost reached.

13.6. Recall the winning strategy of game 7.6. Use the fact that 37 is an odd number.

14.1. Let one of the quadruples be a, b, c, d. Express another quadruple using these numbers. Is it possible to find a third one?

14.2. Find the maximum numbers of vertices of a meandering zigzag that can be covered with n straight lines.

14.3. Try to apply the ideas from the earlier problem with a 4×4 garden (see Problem 6.4).

14.4. *a)* Try to divide the pie in a symmetric way. *b)* Halve each piece obtained previously. *c)* Continue halving.

14.5. First consider numbers only having prime factors ending in 3 or 7.

14.6. Suppose that the initial position is three even heaps. If two smart players play against each other, they will only choose positions with three even heaps competing to reach position $(2, 2, 2)$ first. Find the reason for this.

Solutions

1. Discover the Island of Oxisz

1.1. Embassies on the Island of Oxisz

"I think it is a good idea to put an embassy in country G, since it is centrally located and it has a lot of neighbors. By doing so, we ensure that D, E, F, H, J, L are partners, as you can see on Fig. 57." suggested Albrecht to his peers as they began thinking about the problem (Fig. 58).

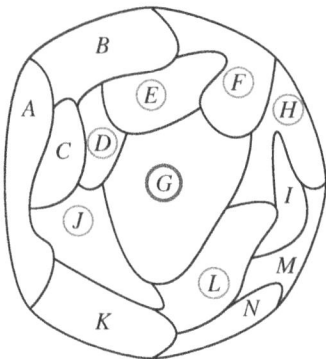

Figure 57

"There are seven countries to be taken care of," said Zsordi, "and one embassy won't suffice. However, putting two embassies, for example in countries A and M we can readily ensure that all countries become a partner."

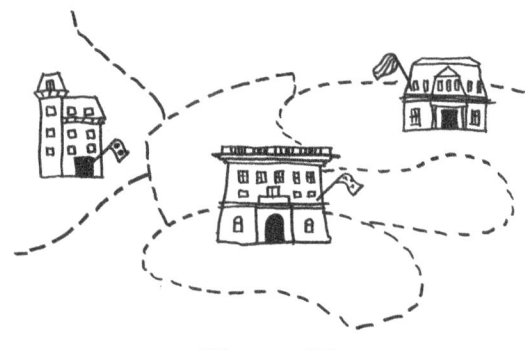

Figure 58

"So we can solve the problem with three embassies, and their placement was chosen sensibly. But are we sure that this is the optimal solution? Or maybe there is an other way to solve the problem with a smaller number of embassies?" asked Albrecht.

The team had barely begun to ponder when Tarkal exclaimed, "Two embassies could be enough!"

Tarkal's solution. Two embassies in countries B and L will make all countries partners of Bergengocia, as it can be verified in the diagram below (Fig. 59).

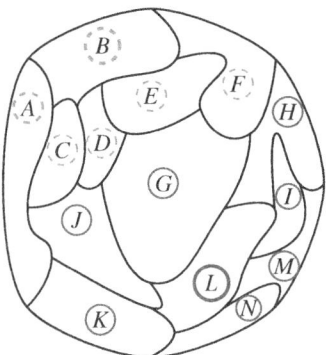

Figure 59

One embassy will not suffice. For example, A and H are not neighbors and do not have a common neighbor, so it is not possible to make them partners using a single embassy.

After solving the problem the team also proved that the only way to get a solution with two embassies is by choosing countries B and L. If we count the number of neighbors for each country, we find that L has 7, G and J have 6, $B, C,$ and D have 5, and the rest have 4 neighbors or less. Apart from the two countries with the embassies, a further 12 countries need a neighbor with an embassy. The only possibilities involve choosing L and one of B, C, D, G, J, or choosing G and J. Checking these six possibilities, it is easy to see that only B and L will provide a solution.[6]

1.2. Zoo Runway

Solution. Complete the diagram to a rectangle, and investigate the horizontal and vertical sides separately (Fig. 60).[7]

---

[6]Increasing the number of countries will make this problem extremely difficult. For example, if there were 100 countries ordered in a haphazard manner, then probably nobody could find the smallest number of embassies needed, not even with the use of a computer.

[7]Architects primarily employ this method to scale the floor plans of buildings.

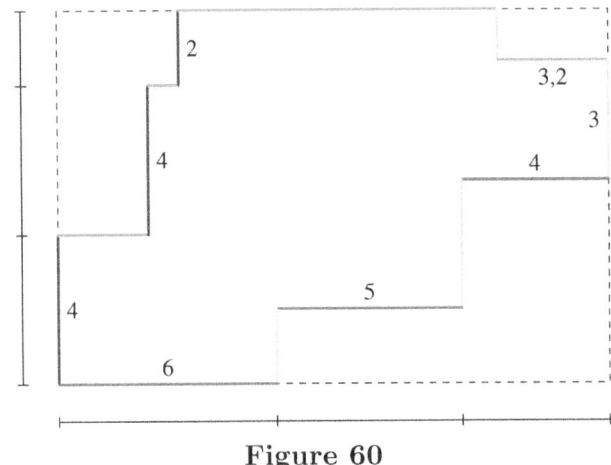

Figure 60

Both the left and right vertical line segments can be combined together to get the shorter side of the rectangle. The total length of the line segments on the left is $4 + 4 + 2 = 10$ m, so the total length of the vertical line segments on the right has to be the same.

Similarly, the total lengths of the horizontal line segments are also the same. The total length of the horizontal line segments on the bottom is $6 + 5 + 4 = 15$ m.

Consequently, the perimeter of the common space is $10+10+15+15 = 50$.

1.3. Incomplete Divisor Sequences

 Solution. The divisors can be divided into pairs according to the picture (Fig. 61).

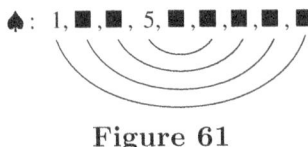

Figure 61

The product of the divisors in each pair is the original number. The divisor in the middle was left without a pair, and so it must be the pair of itself. Thus, the original number is the square of the divisor in the middle.[8]

There are three integer numbers between 1 and 5, and two of them divide the original number. If 4 is a divisor, then so is 2, because numbers divisible by 4 are also divisible by 2. However, if 4 is not a divisor, then only 2 and 3 can be the two divisors.

If the first four divisors are 1, 2, 3, and 5, then the original number is also divisible by 6 (since it is divisible by 2 and 3). In this case the divisor in the middle would be 6, and the original number is its square, 36. However, since 36 is not divisible by 5, this case is not possible.

---

[8]It can be proved in general that a number has an odd number of divisors if and only if it is a perfect square.

If the first four divisors are 1, 2, 4, and 5, then the number is also divisible by 10. Thus, 10 appears in the sequence of divisors.

$$\spadesuit: 1, 2, 4, 5, \blacksquare, \blacksquare, \blacksquare, \blacksquare, \blacksquare$$

Is it possible that 10 appears after the fifth divisor (i.e., the divisor in the middle)? In such a case, the divisor in the middle could be 6, 7, 8, or 9; however, the squares of these numbers are not divisible by 10.

Thus, the divisor in the middle is 10, and consequently, the original number must be 100. The sequence of divisors of 100 does indeed align with the scribbled sequence:

$$100: 1, 2, 4, 5, 10, 20, 25, 50, 100.$$

1.4. Office

Solution. Let us start by examining the value 6. Since there are six people in his row and column, he can see all of them. Let us mark transparent walls with dashed lines and opaque walls with continuous lines (Fig. 62).

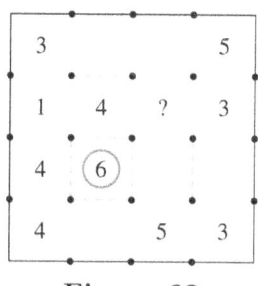

Figure 62

Now, let us examine the two values of 4 in the bottom left part of the image. If there were an opaque wall between them, the circled 4 would only see the other three people in his row, so this case can be eliminated (Fig. 63).

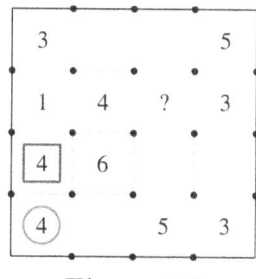

Figure 63

Consequently, the wall between them is transparent. So the framed 4 sees the three people in his row and the circled 4, totaling four people, and thus the wall above him must be opaque.

So the circled 4 can only see the framed 4 above him, and thus he must see everybody in his row (Fig. 64).

54

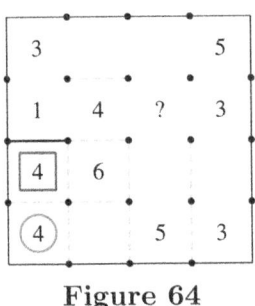

Figure 64

Now consider values 5 and 3 in the bottom right part of the image. Since they can see everybody in their row, we can deduce that they see two and zero colleagues in their columns, respectively (Fig. 65).

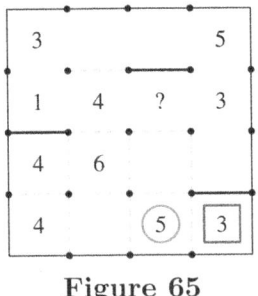

Figure 65

Finally, let us investigate 5 and 3 in the top right part of the image. 5 cannot see 3 in the bottom right, so he has to be able to see everybody else in his row and column (Fig. 66).

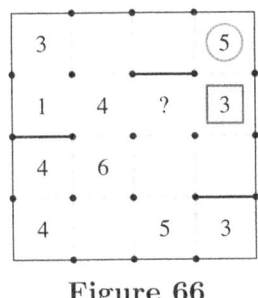

Figure 66

And 3 sees two people in his column, so he will see exactly one co-worker in his row (Fig. 67).

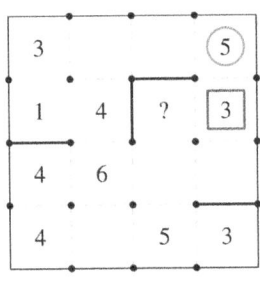

Figure 67

In fact, we have already addressed the problem's question: Litor can see three of his colleagues.

For completeness, let us determine whether the remaining two walls are transparent. To do this, we examine 3 and 1 in the top left part of the image. 3 can only see the people in his row, so he does not see 1, and 1 can only see 4 (Fig. 68).

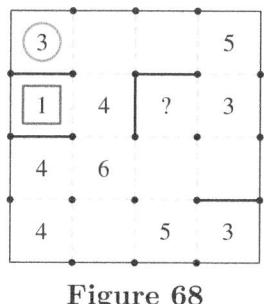

Figure 68

1.5. Bishops and Rooks

 a) The solution of Albrecht. If I place the four rooks in the lower half of the black main diagonal, then the rooks cannot capture each other. It is easy to see that the bishops can only be placed in the top right 4×4 quarter of the board; otherwise some of them could be captured using a rook. Inside this quarter, a bishop cannot be placed on the main diagonal, because it could capture a rook.

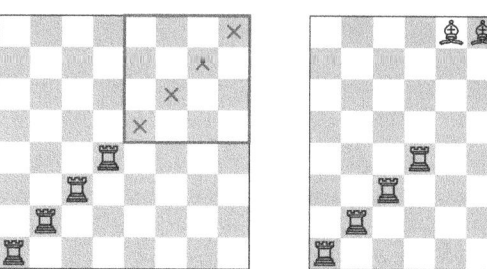

Figure 69

Now, I only have to place the four bishops on the remaining 12 squares so that they will not capture each other, which is way easier than the original problem. The right-hand side of Fig. 69 depicts a possible solution.

Naturally, the problem can be solved in many other ways. For example, Tarkal found a completely different arrangement.

Tarkal's solution. See Fig. 70.

Figure 70

b) Albrecht continued in the following way (Fig. 71):
"Let's try placing the 5 rooks on a main diagonal again:"

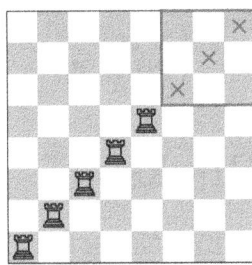

Figure 71

"Now the bishops are confined to a 3×3 part of the board. The remaining part of the main diagonal still cannot be used, so I have to place five bishops on the remaining six squares. However, there will always be two among these that can capture each other. One possible argument is that only one of the four available white squares can remain unoccupied, so there will always be two diagonally neighboring white squares occupied, and the bishops on these can capture each other." "Thus I have proved that it is not possible to place 5 bishops and 5 rooks in a way that no two of them can capture each other."

"This is not entirely correct," interjected Zsordi. "You have only tried one specific arrangement of the 5 rooks. If you choose another, then the squares available for the bishops could be different."

"I did try other arrangements of the 5 rooks, but I have not found one where it is also possible to place the 5 bishops," added Tarkal.

"Then let's arrange the bishops first," suggested Zsordi. "If we make sure to place them in at most 3 different rows and columns (without capturing each other), then the rooks can be arranged in the 5 remaining free rows and columns. Naturally, we still have to make sure that the bishops don't capture the rooks, but I believe it will be possible."

Following Zsordi's advice they found two different solutions for arranging five bishops and five rooks (Fig. 72).

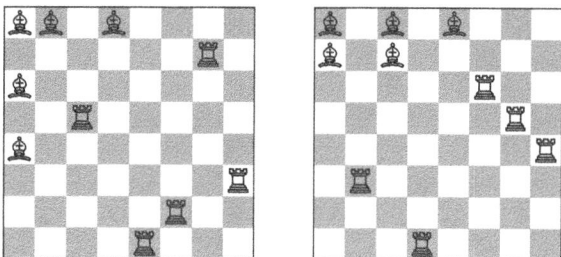

Figure 72

Albrecht felt a bit bad about being wrong. "My argument was indeed incorrect. However, using my method, it can be proved that arranging 6 rooks and 6 bishops on the board according to the rules is impossible. Regardless of how we position 6 rooks in different rows and columns (making sure that they do not capture each other), only 2 rows and 2 columns totalling 4 squares will remain available for the bishops. And placing 6 bishops on 4 squares is clearly impossible."

1.6. Rock–Paper–Scissors with a Twist

In the traditional version of the game we are accustomed to the idea that regardless of the chosen strategy both players have an equal chance winning. Winning the game is based on luck, regardless of the chosen strategy.

However, with the version played on the Island of Oxisz it is a completely different story.

In this version the roles of the players are not symmetric: There will be a first player (who picks the first and third cards from the table) and a second player (who picks the second and fourth cards). The second player has the advantage of seeing the cards picked by the first player and can pick his cards accordingly.

Let us observe a game against Tarkal from the viewpoint of Albrecht (Fig. 73):

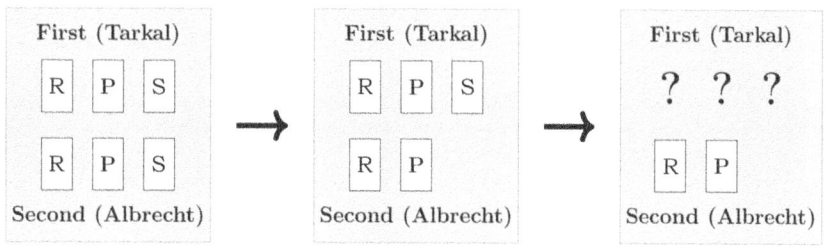

Figure 73

Tarkal started the game, and he took the scissors from Albrecht. At this point, Albrecht argued "What should I take from Tarkal? He took my scissors, so I cannot win against paper. Therefore, I will take paper from him."

Carrying on they successfully proved that this game favors the second player. If the second player is smart, he will surely win, so he has a *winning strategy*.

Solution (winning strategy as the second player). Let us form three pairs from the six cards, which we will refer to as *skew pairs*.

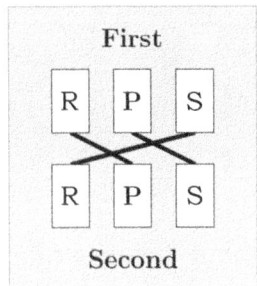

Figure 74

Skew pairs from the point of view of the second player (see Fig. 74):

- The skew pair of our rock is the scissors of the first player.

- The skew pair of our paper is the rock of the first player.

- The skew pair of our scissors is the paper of the first player.

It can be quickly checked that in each skew pair, the card of the second player beats the card of the first player.

We can always win as the second player with the following strategy: We always pick the skew pair of the card that was picked by the first player. Indeed, by following this strategy, two skew pairs will be removed from the table, leaving only the third skew pair. And skew pairs were created with the specific intention that winning the game is equivalent to ending up with a skew pair on the table in the end.

2. Painting with a Boat and a Barrel

2.1. Lifeboat

Upon hearing the problem, Tarkal immediately began solving it: "In each round, we can save 5 people in 3 minutes. In 20 minutes we can fit 6 rounds, and thus we can save $6 \cdot 5 = 30$ people. Done."

"Don't be so hasty," answered Albrecht. "The boat has to return to the ship from the island, which is an additional 3 minutes. Therefore, we have to calculate with 6-minute rounds instead of 3-minute rounds."

"Furthermore, somebody has to row back the boat to the ship," interjected Zsordi. "So in a single round we can only save 4 people."

"You are right, I'm revising my calculations," replied Tarkal. "3 rounds of 6 minutes can be done in 20 minutes, and in each round we save 4 people, so in total we can save $3 \cdot 4 = 12$ people."

"Still wrong," said Albrecht. "It is true that we can get to the ship only three times in 20 minutes, but we can put a fourth batch of people in the boat. These people will only reach the shore at the end of the 21st minute, so only after the ship has sunk, nevertheless they are still going to be saved."

"And in the final round nobody has remain in the boat, therefore, all five people are saved," added Zsordi.

Tarkal took a deep breath and said, "One rower can do a ship-shore-ship round three times, saving 4 people each time. In the end, he takes 4 people from the ship, and rows to the shore, where all 5 people can leave the boat. Therefore, it is possible to save at most $3 \cdot 4 + 5 = 17$ people."

"Now we cannot find any flaws," said Zsordi and Albrecht in unison.

2.2. Barrels of Wine

Zsordi's Solution. If we attempt to use the weight of the empty barrel as the only unknown (denoting it by x), we will not be able to obtain a useful equation using this variable. By also considering the weight of the wine in the full barrel as a second unknown (denoted by y), we immediately obtain two equations:

$$x + y = 188 \quad \text{and} \quad x + \frac{y}{2} = 115.$$

In the first equation the left-hand side is greater by $\frac{y}{2}$, and this is the same as the difference between the right-hand sides: $188 - 115 = 73$. In other words, if we subtract the second equation from the first one, we find that $\frac{y}{2} = 73$.

It is easy to get the answer to this problem from here:

$$h = \left(h + \frac{b}{2}\right) - \frac{b}{2} = 115 - 73 = 42.$$

 Tarkal's Solution. If we empty half of the wine from the full barrel, it gets lighter by $188 - 115 = 73$ kg. If we also empty the other half, we lose another 73 kg; therefore the weight of the resulting empty barrel is $115 - 73 = 42$ kg.

"Interesting," remarked Albrecht. "If I wrote down Tarkal's solution using equations, I would get Zsordi's solution."

2.3. Train Cars

 Solution. We examine the railway cars by type during the solution process. (Figure 75 shows pictograms for each type of car.)

Figure 75

However, if we were to do this in the order they were presented in the problem, the number of conditions to handle would get out of hand. The idea is to investigate cars in the reverse order.

Let us begin with the number of dining cars. Our situation is really simple, since the condition in the problem states that the number of dining cars is at most two.

What can be the number of sleeping cars in a train? Since there must be a dining car between any two sleeping cars, the number of sleeping cars can be at most three.

What can be the number of first class cars in a train? Similarly to what we did in the previous step, it is easy to see that the number of first class cars in a train can be at most four (Fig. 76).

Figure 76

In a similar manner we can deduce that the number of second class cars is at most five. Consequently, the number of cars in a train can be at most $2 + 3 + 4 + 5 = 14$. Is it possible to reach this number?

Yes it is, and we can easily find a solution using our picture. Place the items in the second row between the items of the first one. Now do the same with the third row (making sure that the conditions are still satisfied), and so on. Figure 77 shows one of the many possibilities:

Figure 77

In our attempts to solve the problem, the resulting trains usually satisfy the conditions but contain less than 14 cars.

It is worth verifying that any train meeting the conditions can be completed by adding cars to a total of 14 cars.

2.4. Contemporary Painter

 a) Many different solutions are possible; we believe that the one presented in Fig. 78 is the easiest to verify.

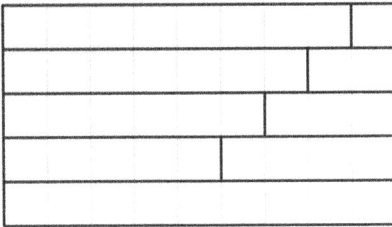

Figure 78

 b) Let us attempt to create a division that meets the conditions of the problem. What kind of rectangles should we pick?

It seems natural to start with small rectangles. Let us examine the ten rectangles that have the smallest areas:

1×1; 1×2; 1×3; 1×4; 2×2; 1×5; 1×6; 2×3; 1×7; 1×8 (vagy 2×4).

The total area of these is
$$1 + 2 + 3 + 4 + 4 + 5 + 6 + 6 + 7 + 8 = 46.$$

We ran into a problem: The total area of the ten smallest rectangles is still greater than the area of the painting ($5 \cdot 9 = 45$), so the rectangles "do not fit into" the painting.

2.5. The Logo of the "Super(stitious)man"

Albrecht, Tarkal, and Zsordi discovered many solutions to the first three questions. We showcase a selection of the shapes they constructed:

a) A triangle and squares—see Fig. 79.

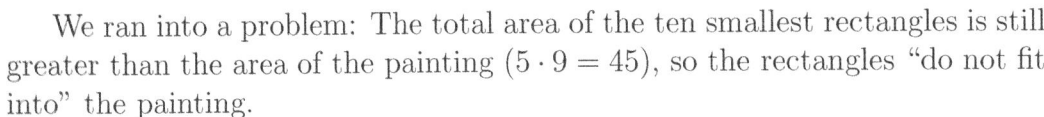

Figure 79

b) An equal number of squares and triangles—see Fig. 80.

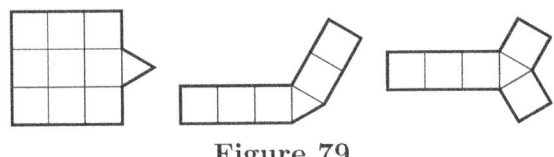

Figure 80

c) Only triangles—see Fig. 81.

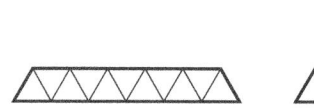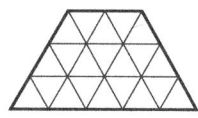

Figure 81

d) Nevertheless, they have not found a single 2D shape composed solely on unit squares with a perimeter of 13 cm. They observed that shapes constructed only from unit squares have an even perimeter. Soon, they have discovered several different proofs for this observation.

 Tarkal's Proof. Let us explore how the perimeter of a 2D shape changes when we add a new square to it. There are four possible cases: The new square can share 1, 2, 3, or 4 sides with the existing squares:

- If it shares one side, then it cancels one side on the perimeter of the original 2D shape and adds three new sides to it. So the perimeter of the 2D shape increases by 3 cm − 1 cm = 2 cm.

- If it shares two sides, then it cancels two sides on the perimeter of the original 2D shape and adds two new sides to it. So the perimeter of the 2D shape does not change.

- If it shares three sides, then it cancels three sides on the perimeter of the original 2D shape and adds one new side to it. So the perimeter of the 2D shape changes by 1 cm − 3 cm = −2 cm; in other words, it decreases by 2 cm.

- If it shares four sides (this is only possible if there is a "hole" in the original 2D shape), it cancels four sides on the perimeter and adds nothing to it. Therefore, the perimeter of the 2D shape decreases by 4 cm.

In all four cases the perimeter of the 2D shape changed by an even number, and the first square has an even number of sides (namely, four), so the perimeter of the 2D shape can only be an even number (measured in cm).

 Albrecht's Proof. Let us imagine a 2D shape consisting only of squares is depicted similarly to what we have seen in the pictures above. Let us mark the external sides using thick line segments and the internal sides (along which the squares were glued together) using thin line segments. The total length of the thick line segments is the perimeter (P) of the 2D shape. If we add twice the total length of the thin line segments (G), we get four times the number of the squares (N):

$$P + 2 \cdot G = 4 \cdot N.$$

The above is true because the thick and thin line segments cover all the sides of the squares in the diagram. Moreover, each thin line segment belongs to exactly two squares, and each thick line segment belongs to exactly one square.

Consequently, the perimeter can be calculated by subtracting twice the total length of the thin line segments from four times the number of the squares, so we subtract an even number from an even number. Thus, the perimeter has to be also even.

Zsordi was inspired by the solution of problem 1.2 and found the following argument.

 Zsordi's Solution. We will form pairs from the sides of the squares that are on the perimeter of the 2D shape. Consider a side a of a square. To find its pair, let us start from the midpoint of this side and go perpendicularly to the side in the direction of the interior of the 2D shape. Eventually, we will reach the perimeter of the 2D shape again at the midpoint of side b of a square. Let us say that side b is the pair of side a (see Fig. 82).

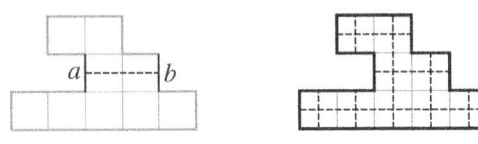

Figure 82

This way we have created pairs from the 1 cm pieces of the perimeter, and therefore the perimeter has to be even.

 Tarkal's Second Proof. If we only use squares, we can imagine our 2D shape as something drawn in a square lattice following lattice lines. The lattice points can be colored black and white alternately (creating a chessboard-like pattern, see Fig. 83). This way every 1 cm piece of the perimeter of the 2D shape connects a black and a white point.

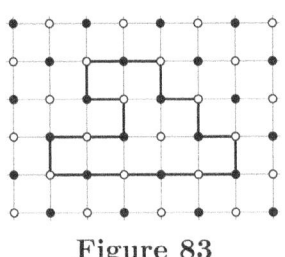

Figure 83

Let us choose one of the vertices of the 2D shape, and let us say it is black. While walking around the perimeter, taking an odd number of unit steps leads us to a white point, whereas an even number of unit steps bring us to a black point. Completing a full circuit brings us back to the initial black point, implying that we have taken an even number of steps. Consequently, the perimeter also has to be even. It is worth noting that if the perimeter consists of multiple disjoint parts, we can apply the same reasoning to each individual part.

2.6. The Game of Covering Numbers

The second player also has a winning strategy in this game. If he plays smart, he will win regardless of his opponent's moves.

Solution (Winning Strategy of the Second Player). As the second player, I win when one of the last two numbers is even, the other is odd.

This can be achieved by ensuring that the number of odd and even numbers remains the same after each of my turns.

The following strategy works: If the first player covers an odd number, I cover an even number; if the first player covers an even number, I cover an odd number.

At this point Tarkal posed the following question. "What if I am the first player, but my opponent is not aware of this strategy? Do I have a chance to win the game? What kind of strategy will give me the best chance to win?"

"Let's assume that your opponent deviated from the winning strategy, and after his turn the number of even numbers is larger than the number of odd numbers," answered Albrecht. "You can win from here as the first player! If you only take odd numbers from this point onward, they will run out before the end of the game. Therefore, two even numbers will remain at the end of the game, ensuring that their sum is also even."

"And what happens if we flip the game?" asked Zsordi. "What if the winning condition is the sum to be even for the second player, and odd for the first player?

"The second player would still have a winning strategy," answered Tarkal. "Furthermore, he would only have to pay attention in the last round. The second player has to cover one of three numbers in the last round. Out of three numbers there will always be two with the same parity (two even or two odd). He should leave these two uncovered, and cover the third one to ensure victory."

3. Skiing Between Fréz and Gard

3.1. Skiing Competition

 Zsordi's Solution. We can keep track the contestants' positions after each slide, see Table 1 (using initials):

Last man sliding	1.	2.	3.	4.	5.	6.	7.	8.
Reichelt	**R**							
Svindal	**S**	R						
Jansrud	S	**J**	R					
Pinturault	S	J	**P**	R				
Küng	S	J	P	**K**	R			
Hirscher	**H**	S	J	P	K	R		
Ligety	H	S	**L**	J	P	K	R	
Miller	H	**M**	S	L	J	P	K	R

Table 1

This is how they stood on the podium: 1st Hirscher; 2nd Miller; 3rd Svindal.[9]

 Tarkal's Solution. To identify the podium finishers, there is no need to find all the intermediate positions during the race. It is sufficient to consider the following:
  a) Hirscher's first place is guaranteed, since after he took the first place, nobody else reached the first position.
  b) Svindal was in the lead until Hirscher achieved a faster time than him; from that point onward Svindal secured the second position.

---

[9] Note that the names used in this problem are the real names of the top skiers from the FIS Alpine Ski World Cup around 2014. This problem was inspired by watching an actual World Cup event on TV.

c) The situation has only changed in the last round, when Miller took the second position from Svindal, who then fell back to the third position (Fig. 84).

Figure 84

3.2. Shopping Carts

 Zsordi's Solution. Let x denote the length of a single cart and y the length of the protruding part of the next cart (measured in meters).

If seven carts are pushed together, the length of the line equals the length of the first cart plus six times the length of the protruding part:

$$3 = x + 6y.$$

If ten carts are pushed together, the length of the line equals the length of the first cart plus nine times the length of the protruding part:

$$4 = x + 9y.$$

Subtracting the first equation from the second one, we get $1 = 3y$. Substituting this back in the first equation, we get $x = 1$, so the length of a single shopping cart is exactly 1 meter.

 Tarkal's Solution. If we pull out the last three carts from a line of ten carts, seven carts remain, and therefore the line gets shorter by 1 meter. If we pull out another three carts from the remaining seven carts, the line will get shorter by 1 meter again, so four carts will have a length of 2 meters. Finally, pulling out another three carts, we end up with a single cart and a length of 1 meter.

"We were lucky with the given data," noted Albrecht after listening to Tarkal's solution. "Let k and n denote the number of carts in the shorter and the longer lines, respectively. Tarkal's method will work when $k-1$ is divisible by $n-k$."

3.3. Connecting a Grid into a Polygon

 Solution. Every polygon has the same number of sides and vertices. Since each of the vertices has to be one of the 20 points in the lattice and each point can be a vertex only once, the polygon can have at most 20 sides.

It is trickier to decide whether we can reach this maximum. We have to find a path that visits all the points in the lattice and also changes direction at each of these points. This is actually possible, see Fig. 85.

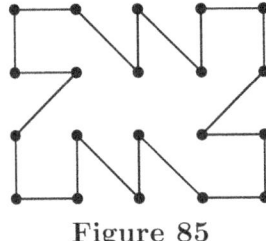

Figure 85

3.4. Buses Connecting Villages

Albrecht looked at the map and cried out Bus no. 41 travels through 5 villages, and there are many-many ways to find a round trip through 5 villages on this map. How will we find the one that will work?"

"We only have to draw," said Zsordi, and she began to sketch a solution to the problem.

 Zsordi's Argument. Let us draw a new map by first sketching the route of bus no. 41 (see Fig. 86). We denote the villages by their initials.

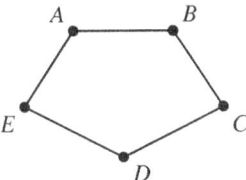

Figure 86

Let us add the villages visited by bus no. 42 and also the connections between them (Fig. 87).

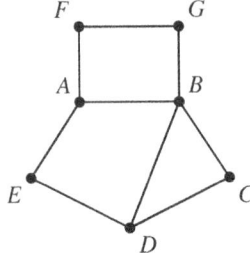

Figure 87

Now add the route of bus no. 43, and compare our map (Fig. 88) with the original one with numbers (Fig. 89).

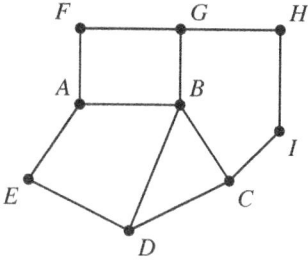

Figure 88

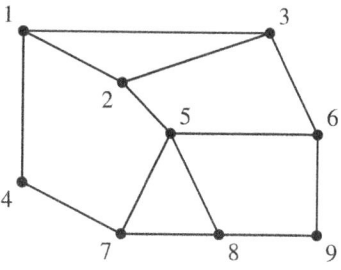

Figure 89

"Although there are some similarities between the two maps, it is still not clear how can we make a one-to-one correspondence between them. For example, B, C and D form a triangle, but we can find two triangles in the picture with the numbers," said Tarkal.

Finally Albrecht managed to establish a one-to-one correspondence between the two maps.

Albrecht's Solution. In the map of Zsordi, B is connected to four villages (A, C, D, and G); thus at least[10] four roads have to connect to village B in the map with the numbers. In the map with the numbers, every village has at most three roads except for village 5; therefore village 5 has to be B (shorthand notation for this will be $B = 5$).

Based on Tarkal's observation about the triangle formed by villages B, C, and D, C and D have to be 7 and 8. Now we have to consider two cases:

- Case 1: $C = 7$ and $D = 8$ (Fig. 90).

- Case 2: $C = 8$ and $D = 7$ (Fig. 91).

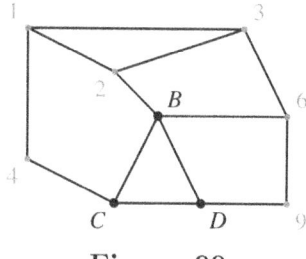

Figure 90

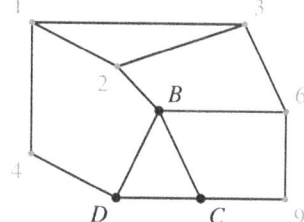

Figure 91

Tracing the route of bus no. 41 on the map, we can note that after segment $B - C - D$ it can continue in a unique manner to E and subsequently to A. Thus in Case 1 $E = 9$ and $A = 6$ (Fig. 92), while in Case 2 $E = 4$ and $A = 1$. But we have to be able to get back to B directly, which is impossible in Case 2 (Fig. 93).

---

[10] At least, because there can be additional segments not used by buses. Actually there must be such a segment: In Zsordi's picture, we can see 12 segments, and in the picture with the numbers there are 13 segments.

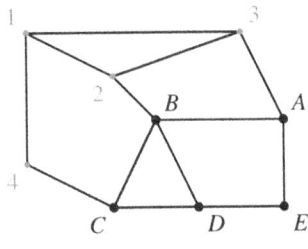

Figure 92

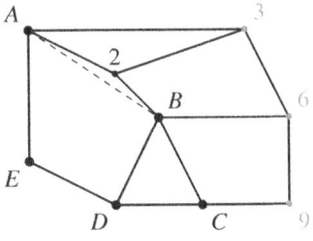

Figure 93

Therefore we only have to deal with Case 1. There is a road leading from B to G; thus $G = 2$. Bus no. 42 only stops in F between A and G, so $F = 3$. Finally, following the route of bus no. 43, we get $H = 1$ and $I = 4$ (Fig. 94).

Thus the villages can be arranged on the map only in the following manner (Fig. 95):

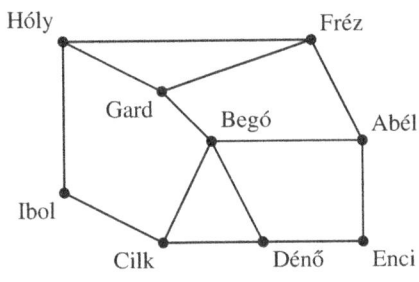

Figure 94

Figure 95

3.5. Who Is My Bigger Neighbor?

a) "This problem is too easy," said Tarkal. "It is enough to arrange the numbers in increasing order from 1 to 9 (Fig. 96)."

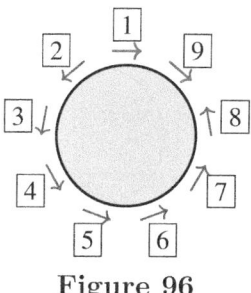

Figure 96

"You are wrong. If you take a closer look, you can notice that the child with card number 1 and the child with card number 9 will both point to the left," answered Albrecht.

"In that case there always will be at least two children pointing to the left, because the right neighbour of the child with number 1 will definitely point to the left, just like the left neighbour of the child with number 9 (Fig. 97)."

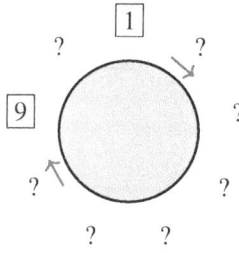

Figure 97

"This is also wrong," said Zsordi. "These two children pointing to the left can actually be the same person, if there is a child whose left neighbour holds card number 9 and right neighbour holds card number 1 (Fig. 98)."

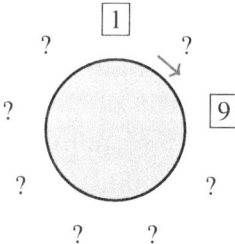

Figure 98

After making these observations, they quickly found a solution to the problem.

 a) In a good arrangement of the cards, the left neighbor of the child with number 1 (let us call him X) will definitely point to the left. What can we say about the left neighbor of the child with number 2 (let us call him Y)? (Fig. 99).

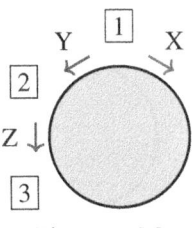

Figure 99

Y cannot coincide with X, so he has to point to the right. His right neighbor has number 2, so he can only point to the right if his left neighbor holds card number 1. As a result, the child sitting two places to the right of the one holding card number 1 must have card number 2.

A similar argument about observing the left neighbor of the child holding card number 3 (let us call him Z) shows that card number 3 has to be placed two places to the right from card number 2. Continuing this argument we arrive at the following arrangement of the cards (Fig. 100):

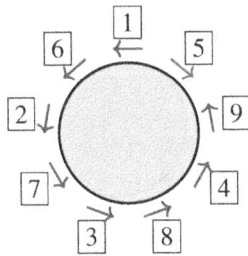

Figure 100

Only the left neighbor of the child with card number 1, namely, the child with card number 5, points to the left; everybody else points to the right.

b) When the team tackled the case with eight children, they discovered that the ideas from the first part still worked for a while.

 Solution. Number 2 has to be two places to the right of number 1, number 3 has to be two places to the right of number 2, number 4 has to be two places to the right of number 3, and number 5 has to be two places to the right of number 4. But hold on a second, and we already have number 1 there (Fig. 101).

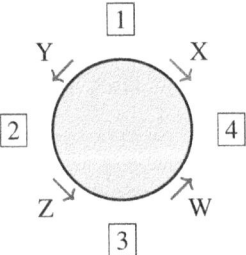

Figure 101

This is not a problem in itself, because so far only the child between number 1 and number 4 has to point to the left. However, we have to give number 5 to somebody: At this point only X, Y, Z, or W can get it. No matter which of

73

them gets the card, his right neighbor will also point to the left, because his left neighbor can only have card number 6, 7, or 8.

After the solution, Albrecht mused:

"Our solutions also show that there is always a good distribution for an odd number of children (and this distribution is unique up to a rotation), while there is no such distribution for an even number of children."

Albrecht's Argument for Even Number of Children. Let us provide each child with a second card, alternating between letters A and B.

Let us find the child with the largest number among the children holding a card with letter A. His right neighbor has to point to the left, because both of his neighbors hold a card with letter A, and his left neighbor has the biggest number among all the children holding a card with letter A.

Similarly, let us find the child with holding the largest number among the children holding a card with letter B. His right neighbor also has to point to the left, because both of his neighbors hold a card with letter B, and his left neighbor holds the biggest number among all the children holding a card with letter B (Fig. 102).

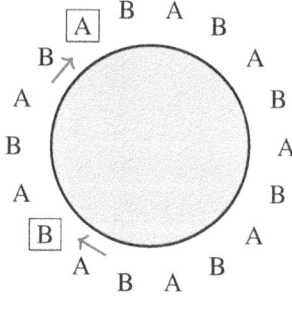

Figure 102

So we have found two different children pointing to the left, and thus the conditions of the problem cannot be satisfied.

3.6. Placing Tokens in Four Squares

In this game the first player has a winning strategy. He can play in a way that the second player has no chance to win.

Solution (Winning Strategy for the First Player). The second player's goal is to have a different number of tokens in all four squares. This is only possible if the squares contain 0, 1, 2, and 3 tokens (not necessarily in this order). This implies that the second player will inevitably lose if either of the following two conditions arises during the game:

- There is a square with at least four tokens.

- Each square contains a token.

Playing as the first player we want either of these two conditions to arise.

After placing our first token, the second player has essentially two choices:

1. He can choose the same square where we placed our token, resulting in a square with two tokens. If we place our remaining two tokens in the same square, it will contain at least four tokens securing a win for us.

2. If the second player chooses a different square, there will be two squares with one token each. If we place our remaining two tokens in the remaining two empty squares, each square will contain at least one token. So we have won.

4. Football in Brassó

4.1 Stage-coaches

Pozsony and Brassó are not situated on the island of Oxisz; instead, they are genuine cities in East-Central Europe. This problem is excerpted from the novel "Strange Marriage" by the renowned Hungarian writer Kálmán Mikszáth.[11] In the tale, the problem is presented by a Hungarian nobleman, István Horváth, to a successful businessman from Brünn, who desires to marry the beautiful Rozália, Horváth's daughter. A correct solution to this puzzle is a prerequisite for any potential suitor. As the story unfolds, we discover the solution to the math problem and the result of the marriage proposal as well.

Solution of Kálmán Mikszáth. "Twenty."

"Now, now, young man. Think again. You are wrong."

"Perhaps there is a catch in that question?"

"I assure you, no."

"In that case, the number will undoubtedly be twenty. Or rather, one coach, the twentieth, will have got to Pozsony by the time I am leaving, which leaves only nineteen. Is that right?"

"No, no. I am very sorry, young man."

"Please permit me to work this out quietly in my room."

"You're welcome," the old man said, smiling scornfully.

The business man from Brünn spent the whole day over his calculations; he worked in a frenzy, filling some ten sheets of paper with figures. He worked at the problem till the perspiration ran from his brow, yet he could not find the solution, for after each calculation he arrived at a different number. At last he asked the cook to give him a basket of beans, and these he laid out to represent the coaches so that he could visualize their movements, but the beans only added to his confusion.

When Horváth saw that the suitor could not find the answer, he relieved him of his uncertainty.

---

[11] The novel was originally written in 1900 with the Hungarian title "Különös házasság." It was translated into English by István Farkas and published by Corvina Budapest in 1964; we are taking extracts from this edition.

"It is clear that you will never make a good business man, because you can't see things as they are really happening. During the ten days of your journey on the Pozsony-Brassó road you would find the carriages that had left during the previous ten days. Accordingly, there are forty coaches on the Pozsony-Brassó road. As for your own road ahead, I can see on it – and pray, do not take this amiss – nothing but a mitten."

Upon reading the excerpt from the novel, Tarkal began shaking his head:

"Let's suppose," he mused, "that we set off from Pozsony on May 11th. If both carriages departed from Brassó early on May 1st, thus arriving in Pozsony before our departure, then we won't encounter them anymore. We reach Brassó on May 21st. If the mail coaches of that day depart from Brassó only after our arrival, then we won't encounter those either. Hence, we only come across the carriages that departed from Brassó between May 2nd and May 20th, spanning 19 days, which amounts to a total of 38 mail coaches."

Zsordi further elaborated on the train of thought:

"By selecting appropriate departure and arrival times within that timeframe, the total count could be 39, 41, or possibly even 42."

Tarkal continued:

"Even if we were to hypothesize that the mail coaches depart at the same hour daily and that the journey consistently lasts precisely 240 hours, even then we should have to tinker with how we calculate if a carriage starts at the exact moment another one arrives."

"You've observed it correctly, but what's certain is that it's definitely not 20 (nor is it 19) as Rozália's suitor claimed, and that's enlightening in itself. My suggestion is that when assigning this task to someone, we should consider any answer between 38 and 42, along with a thorough explanation," concluded Albrecht, putting an end to the discussion.

"We can all agree on that."

And for those concerned about Rozália's fate in the novel, rest assured: Later on, a suitor, named Bezerédj from the Transdanubia region, distinguished by his impressively long mustache, not only answered this question but also tackled the subsequent ones, hence ultimately "he won his princess."

Remark on the history of Pozsony, Brassó, and the term "coach." Pozsony is the Hungarian name of Bratislava, which has been the capital city of Slovakia since 1993. Brassó is located in Transylvania, which became part of Romania in 1920; the Romanian name of the city is Braşov.

Prior to 1920, both cities belonged to the Hungarian Kingdom (which was the part of the Austro-Hungarian Empire, whose map is depicted in Fig. 103); hence, their Hungarian names are used in Mikszáth's novel and the problem.

However, they are also frequently referred to by their German names: Pressburg and Kronstadt, respectively, due to their historical association with German-speaking craftsmen and merchants. Brünn, the hometown of the businessman deceived by István Horváth, was another city within the Austro-Hungarian Empire. Today, Brünn belongs to the Czech Republic and is known by its Czech name, Brno.

The route from Pozsony to Brassó passes through a village named Kocs, which sounds like "coach" in Hungarian. Stage-coaches often made stops there to change horses. In the fifteenth century, the renowned carriage-makers of this village developed a faster and lighter vehicle that quickly gained popularity across Europe. Consequently, the English term "coach" and its equivalents in German ("Kutsche") and Spanish ("coche") have their origins in the name of this Hungarian village.

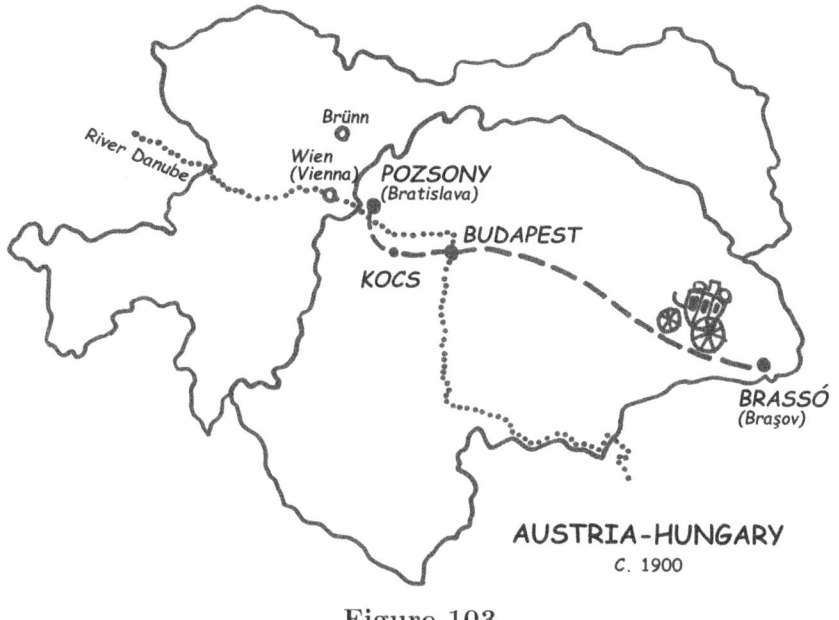

Figure 103

4.2. Pyramid of Numbers

 Solution. First observe the part in the frame in Fig. 104. Since numbers 16 and 20 are close to each other, it is reasonable to expect that the two numbers next to 16 are small, and so we will be able to determine them.

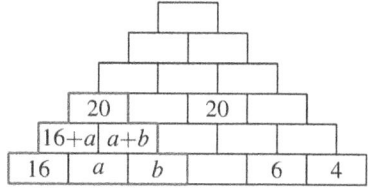

Figure 104

Let us denote the two numbers in the bottom row by a and b. All the numbers are at least 1. If a and b were 1, then the first two sums in the row above them would be 17 and 2. However, this is not possible, because the sum of these two numbers is 19, not 20. Therefore, a and b cannot be 1. If one of them is 2, then the second element of the second row is 3, so the first element can only be 17. Thus the numbers in the bottom row are 16, 1, and 2.

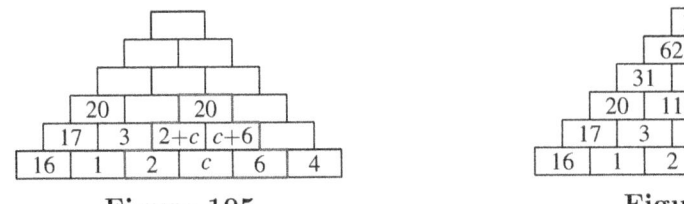

Figure 105 **Figure 106**

Now let us investigate the area below the other 20 in the pyramid (Fig. 105).

Denote the middle number by c. Express 20 as the sum of the numbers below it: $20 = (2+c)+(c+6) = 8+2c$, so $c = 6$. This way we have determined all the numbers in the bottom row, and we can easily find the numbers that are still missing, see Fig. 106. The solution is uniquely determined.

4.3. Photocopied Polygon

Solution. All the lengths of the sides of the 9-gon can be determined by comparing them to either the shortest or the longest side.

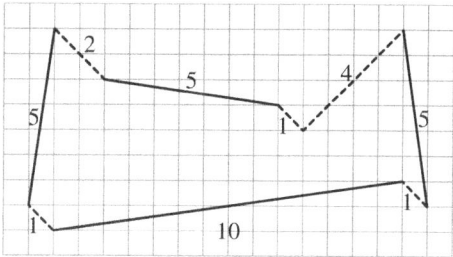

Figure 107

See Fig. 107. The three shortest sides among those marked with dashed lines are congruent: Each of them is a diagonal of a unit square in the square lattice. Thus the lengths of these are 1 cm. The other dashed sides are a combination of 2 and 4 such diagonals, respectively.

Around the line segments drawn with a solid line in Fig. 107, we can draw congruent rectangles with seven units on the longer side and one unit on the shorter side, as in Fig. 108.

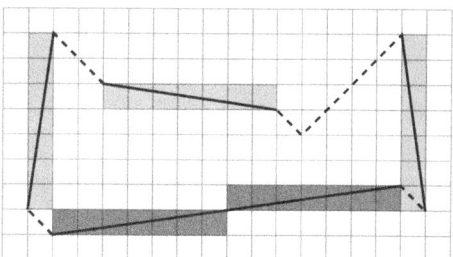

Figure 108

Since the longest side (10 cm according to the text of the problem) consists of two such diagonals, one diagonal is just 5 cm long.

By adding up the lengths of the sides of the 9-gon, we can obtain its perimeter:

$$1 + 10 + 1 + 5 + 4 + 1 + 5 + 2 + 5 = 34 \text{ cm}.$$

In our solution first we determined the relative lengths of the dashed lines and then the relative lengths of the solid lines. However, it is actually possible to find a connection between these two types of sides. The diagonal of the 7×1 rectangles in the grid is five times as long as the diagonals of the unit squares.

This can be proved using the Pythagorean theorem. If the unit in the lattice is x, then the diagonal of a unit square is $\sqrt{x^2 + x^2} = \sqrt{2}x$, while the diagonal of a 7×1 rectangle is

$$\sqrt{(7x)^2 + x^2} = \sqrt{50x^2} = 5\sqrt{2}x.$$

This means that in fact it would have been enough to reveal the length of the shortest side of the 9-gon only, because all the lengths of the sides and the perimeter can be determined from that information. Naturally, this would have made the problem harder.

4.4. Polygon with Many Concave Angles

In a convex n-gon draw, all the diagonals possible from one vertex, thus dividing the polygon into $n - 2$ nonoverlapping triangles. The sum of the interior angles of all the triangles is the same as the sum of the interior angles of the n-gon: $(n - 2) \cdot 180°$. The proof is not so simple for concave polygons, but the statement is also true. During the solution we will take advantage of this fact.

Tarkal's Solution. Let us use the hint and count how many concave angles "fit into" the polygon.

The sum of the internal angles in a 12-gon is $10 \cdot 180°$. A concave angle is greater than $180°$. This means that the polygon cannot have ten (or more than ten) concave angles, because the sum of these would be greater than $10 \cdot 180°$.

So at most nine concave angles "fit into" the polygon, and we show an example that this is possible (Fig. 109).

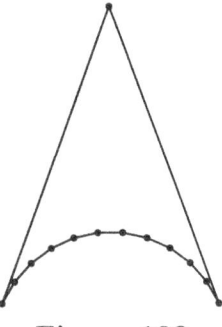

Figure 109

"How did you find this construction?" asked Albrecht.

"First I estimated the approximate size of the concave angles. It came out that the average of the concave angles has to be smaller than $\frac{10}{9} \cdot 180°$. So I had to find a polygon with concave angles that are just a little bit bigger than $180°$. After that I quickly found this polygon," Tarkal explained.

"I think the solution is similar to the 2.4. problem, where we helped the contemporary painter," suggested Zsordi.

"What do you mean exactly?" asked Tarkal.

"When we wanted to divide the painting into 10 different rectangles, we have found that the rectangles 'didn't fit' into the painting (because their total area was larger than the area of painting)," Zsordi explained.

"This is an interesting connection," agreed Albrecht. "In both cases we used quantities with which we could do calculations, and then compared the results."

4.5. Mixed Up Kits

The hikers approached part *a)* in the following way.

"I think we should start by determining the 11 sums," suggested Zsordi. "After that, it will surely be easier to find the appropriate pairs."

"The sum on the jerseys is $1 + 2 + 3 + \cdots + 11$, which is 66," answered Tarkal quickly. "The sum on the shorts is the same, making the total 132."

"Now we only have to find 11 consecutive numbers with a sum of 132," suggested Zsordi.

"I've found it," said Tarkal after a few minutes. "I checked the sum of 11 consecutive numbers starting from 1, then starting from 2, and so on. When the smallest number was 7, I got $7 + 8 + 9 + \cdots + 17 = 132$, which is exactly the sum we needed."

"Now let's try to pair up the jerseys and the shorts accordingly," said Albrecht.

"I suggest assigning jersey no. 1 and shorts no. 1 to the two smallest sums, 7 and 8," suggested Zsordi. "Let's do the same with the two 2s by assigning them to sums 9 and 10, and so on."

Following this method, they obtained pairs described by Table 2.

Jersey	6	1	7	2	8	3	9	4	10	5	11
Shorts	1	7	2	8	3	9	4	10	5	11	6
Sum	7	8	9	10	11	12	13	14	15	16	17

Table 2

"Can you guess what this reminds me of?" wondered Zsordi. "In problem 3.5 the distribution of the cards was similar. It seems like we used the same arrangement on the jerseys and shorts, just the numbers on one of them were shifted by one."

"This is a very interesting observation," said Albrecht.[12]

The team continued with part *b)*.

"This can be solved just the same way as part *a)*, right?" asked Tarkal.

 Tarkal's Solution to Part b). Now, the total of the numbers on the jerseys and the shorts is $2 \cdot (2 + 3 + \cdots + 11) = 2 \cdot 65 = 130$. Therefore, we need to find ten consecutive numbers with a sum of 130.

I started adding up ten consecutive numbers again. If we begin from 1, the sum is $1 + 2 + 3 + \cdots + 10 = 55$. If we start from 2, it is $2 + 3 + 4 + \cdots + 11 = 65$, and so on. When we start from 8, the sum is $8 + 9 + \cdots + 17 = 125$, which is less than 130; however, starting from 9, the sum is $9 + 10 + \cdots + 18 = 135$, which is too much.

And if we start from an even bigger number, the sum would be even bigger. Consequently, it is not possible to find ten consecutive numbers with a sum of 130, so it is not possible pair up the jerseys and the shorts in this case.

---

[12] The connection is not a matter of chance, yet it is not entirely self-evident either. Filling in the missing details is left to the reader.

"I cannot add up numbers as fast as Tarkal, however, I think I have a different approach," said Zsordi.

"Tell your approach first, and then I will happily show you how I do it," said Tarkal.

 Zsordi's Solution for Part b). "Among ten consecutive numbers there are always exactly 5 odd and 5 even numbers," began Zsordi. "Therefore their sum will always be odd. However, 130 is even, so it cannot be the sum of 10 consecutive numbers."

"Wow, this approach is truly elegant," said Tarkal. "I'm feeling a bit guilty about all the calculations in my solution."

"Can you still tell me how did you add up so many numbers so quickly?" asked Zsordi.

"So many *consecutive* numbers," added Tarkal. "That's the key to my method. Consider, for instance, sum $7 + 8 + \cdots + 17 = X$. Now, write down the same addends in reverse order beneath it." (See Table 3.)

$$
\begin{array}{ccccccccccc}
7 & + & 8 & + & 9 & + & \ldots & + & 16 & + & 17 & = & X \\
17 & + & 16 & + & 15 & + & \ldots & + & 8 & + & 7 & = & X \\
\hline
24 & + & 24 & + & 24 & + & \ldots & + & 24 & + & 24 & = & 2X
\end{array}
$$

Table 3

The sum is 24 in every column, so we have to add these up a few times. "How many times exactly?" asked Tarkal from Zsordi.

"11 times, correct?" asked Zsordi, sounding a bit uncertain.

"Exactly," agreed Tarkal. "An easy mistake to make is to say that we added $17 - 7 = 10$ numbers, but this is wrong! In fact, there are 11 numbers, since if 7 is the 1st, 8 is the 2nd, 9 is the 3rd, then we can observe that the difference of the two numbers is always 6, thus 17 is the 11th number."

"Or we can recall that the problem is about 11 players," laughed Zsordi.

"Indeed, in this specific case we can recall this, but we have to pay attention to what I said in the general case in order to avoid mistakes," said Tarkal. Thus we can see $11 \cdot 24 = 2X$, and therefore

$$7 + 8 + \cdots + 17 = X = \frac{11 \cdot 24}{2} = 132.$$

"This method is called Gauss' trick,"[13] added Albrecht. "And it can also be used when the addends are not consecutive numbers. For example, can you tell me the sum of the odd numbers from 1 to 99?"

"I'll try this," said Zsordi "Write down the sum $N = 1+3+5+\cdots+97+99$, and then the same addends in the reverse order beneath it." (See Table 4.)

---

[13]Many may be familiar with the story of the schoolboy Gauss surprising his teacher by quickly finding the sum of first 100 numbers using the method described above. While we do not have any proof that this has actually happened, the name for the method is widely recognized, and we will also refer to it that way.

$$\begin{array}{ccccccccccc}
1 & + & 3 & + & 5 & + & \ldots & + & 97 & + & 99 & = & N \\
99 & + & 97 & + & 95 & + & \ldots & + & 3 & + & 1 & = & N \\
\hline
100 & + & 100 & + & 100 & + & \ldots & + & 100 & + & 100 & = & 2N
\end{array}$$

Table 4

Again, the sum of the numbers in each column will be the same. We only have to find out how many numbers did we add up. From 1 to 100 there are 100 numbers, and half of these are odd, so we have a sum with 50 addends. Based on this, $50 \cdot 100 = 2N$; therefore

$$1 + 3 + 5 + \cdots + 97 + 99 = N = 50^2 = 2500.$$

"Exactly," agreed Albrecht. "In a similar way it is possible to prove that the sum of the odd numbers from 1 to $2n - 1$ equals n^2."

Working out the details is left to the reader.

4.6. Placing Tokens on a Board with five Squares

We will describe the game positions using 5-tuples (similar to Albrecht and his team).

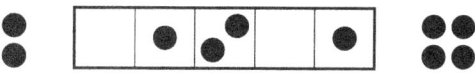

Figure 110

For example, the position in Fig. 110 is represented by 5-tuple $(2, 1, 1, 0, 0)$, meaning that there is a square with exactly two tokens, there are two squares with one token each, and there are two empty squares. The order of the squares is not important in this game, so we can always choose a decreasing sequence of numbers in these 5-tuples.

"What do you think, the first or the second player has a winning strategy?" asked Zsordi.

"The second player will place twice as many tokens, and also makes the last move in the game. This seems to be advantageous," answered Tarkal.

"Although this is true, the second player can only win by reaching position $(4, 3, 2, 1, 0)$ in the end," answered Albrecht. The reason for this is that this is the only way to distribute ten tokens in five squares in a way that each square has a different number of tokens.

"So if five tokens end up in the same square, the first player's win is guaranteed. The same is true if each square contains at least 1 token. So the second player has to steer clear of these ," said Zsordi. A similar idea was used in the simpler version of the game: 3.6. This also explains why it is not surprising that Zsordi's observation is once again useful for finding the strategy.

 Solution. 1. At the start of the game, the first player essentially has two choices: He can place his token next to the token already placed, or he can place it in an empty square, thus reaching position $(2, 0, 0, 0, 0)$ or $(1, 1, 0, 0, 0)$.

2. Now it is the second player's turn, and after his move, the number of tokens on the board will be 4. If at least three of the four tokens are on the same square, the first player could win the game, because he still holds two tokens. If he places both his tokens in the square with at least three tokens, then there will be at least five tokens on this square at the end of the game.

   A very similar idea works in the case when there are tokens on at least three distinct squares. The first player could put his last two tokens on the remaining two squares, so there would be a token in each square in the end.

   So the second player has to ensure that the four tokens are distributed between two squares, and no square contains at least three tokens. There is only one such position: $(2, 2, 0, 0, 0)$. Therefore, the only chance for the second player to win the game is to create this position using his two tokens. It is easy to check that he can always achieve this.

3. Let us see how the first player reacts. Using a single token he can reach either position $(2, 2, 1, 0, 0)$ or position $(3, 2, 0, 0, 0)$.

4. Now it is the second player's turn. Since the first player still has a token, the second player should avoid creating a square with at least four tokens, and he also has to make sure that there are no four non-empty squares. Seven tokens can be arranged in two ways to achieve this: $(3, 3, 1, 0, 0)$ and $(3, 2, 2, 0, 0)$. Luckily for the second player, these positions can always be reached after the first player's second move.

5. It is the last move of the first player. If the second player created position $(3, 3, 1, 0, 0)$, then the first player can reach positions $(4, 3, 1, 0, 0)$, $(3, 3, 2, 0, 0)$, and $(3, 3, 1, 1, 0)$ by placing his last token. From position $(3, 2, 2, 0, 0)$, he can also reach position $(3, 2, 2, 1, 0)$.

6. From all four positions, the first player can reach after his last move, and the second player can reach the desired position of $(4, 3, 2, 1, 0)$ using his last two tokens.

Therefore, the second player has a winning strategy: If he plays well, he will surely win. Figure 111 summarizes a game where the first player makes no mistake:

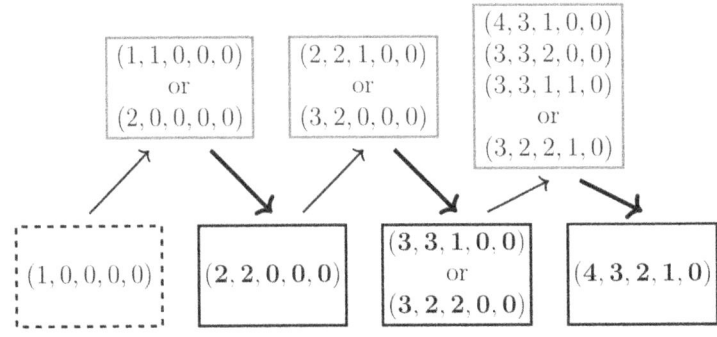

Figure 111

The argument also shows that a single mistake from the second player gives a chance for the first player to win the game, because he can either guarantee at least five tokens on the same square or at least five non-empty squares.

5. Can I Lock the Cat in the Mill?

5.1. Magic?

"I was born on the 23rd," said Tarkal. "These were my results: 23, 40, 80, 72, 9, 36, 42. Zsordi, what numbers have you got?" "I was born on the 27th, and my results were 27, 44, 88, 72, 9, 36, and finally 42."

"Look, starting from 72 both of us have the same results. I wonder whether these will always be the last four numbers," said Tarkal.

"Well, not for me," said Albrecht. "I was born on the 21st, and my results were 21, 38, 76, 63, 9, 36, 42."

"The last four numbers are not the same indeed," admitted Tarkal. "But your last three numbers are also 9, 36, 42. My guess is that this will be true regardless of the starting number."

"I agree, though I also have an additional observation," replied Albrecht. "The number before 9 is 72 for you, 63 for me, and both are divisible by 9."

"And is this something important?" asked Zsordi.

"Yes, this is the actual basis of the trick," answered Albrecht.

Solution. The initial number has to be between 1 and 31. After adding 17 the result will be between 18 and 48. After multiplying by two, the result will be between 36 and 96. The magic happens at this point.

When we subtract the sum of the digits from a number with two digits, we get a number divisible by 9, as the team noticed. What is the reason? If our number has the form $\overline{ab}$,[14] subtracting the last digit the result is $\overline{a0}$, i.e., it is ten times the first digit. Now, if we subtract the first digit, we get $10 \cdot a - a = 9 \cdot a$.

So after this step the result is a two-digit number divisible by 9, and we have to add up the digits of this number. We claim that this will always be 9, as Tarkal suspected.

---

[14] $\overline{ab}$ is the usual notation for a number whose first digit is a and the second digit is b. Another way to write it is $10a + b$, but $\overline{ab}$ refers more directly to the base ten representation of the number. The line above the number is used to differentiate it from the product of a and b which is written as ab.

We know that the sum of the digits of a number that is divisible by 9 is also divisible by 9, so it has to be 9, 18, 27, etc. But in our case it can only be 9, since 99 is the smallest number where the sum of the digits is 18, and we have already seen that our result at this point will be smaller than 96.

Therefore the current result in the notebook must be 9.

Multiplying by 4, we get 36. Finally, adding 6 will result in 42.[15]

"I like this problem. I'll show it to my friends at home, maybe they will think I'm a psychic," Zsordi laughed.

5.2. Tortoise, Cat, and Table

 Zsordi's Solution. Let us imagine having copies of both the table and the cat. Now, let us stack the two tables on top of each other. Place one cat on the ground, the tortoise on the top of the lower table, and the other cat on the top of the upper table, see Fig. 112.

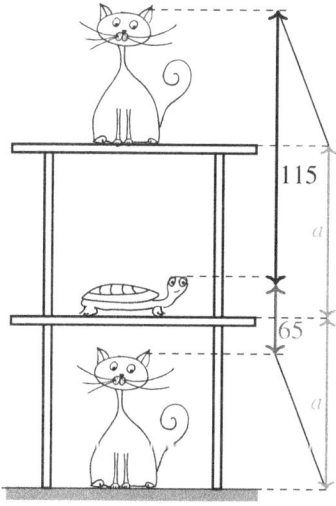

Figure 112

The head of the tortoise is 65 cm above the head of the lower cat, and the head of the upper cat is 115 cm above the head of the tortoise. Therefore, the difference between the two cats is $65 + 115 = 180$ cm. But the difference between the two cats is exactly the height of the two tables, so the height of a single table is $\frac{180}{2} = 90$ cm.

 Albrecht's Solution. The problem can also be solved by setting up a system of equations. Let c and t denote the height of the cat and the tortoise, respectively. Let h denote the height of the table (everything is measured in centimeters).

---

[15] The same as the answer to the question about life, the universe, and everything in "The Hitchhiker's Guide to the Galaxy" by Douglas Adams. This book is very popular among some organizers of the Dürer Competition. Probably not independently from this, we often propose problems where the answer is 42.

From the information provided in the problem, we derive the following equations:

$$h + c = t + 115$$
$$h + t = c + 65.$$

Adding up these two equations, we get

$$h + c + h + t = t + 115 + c + 65.$$

After collecting the like terms and subtracting $(c + t)$ from both sides, we get

$$2h = 180,$$

thus $h = 90$, and therefore the height of the table is 90 cm.

Tarkal observed that the trickiest step in the solution of Albrecht (adding up the two equations) is analogous to Zsordi's beautiful idea of putting two tables on the top of each other.

5.3. Central Lock

 Tarkal's Solution. After waiting a minute, the lock will be either in state (T) or in state (C). If we push the button twice, the lock will be either in state (O) or state (T), so waiting another minute it will definitely be in state (T). Therefore, pushing the button a third time, we reach state (C) (Fig. 113).

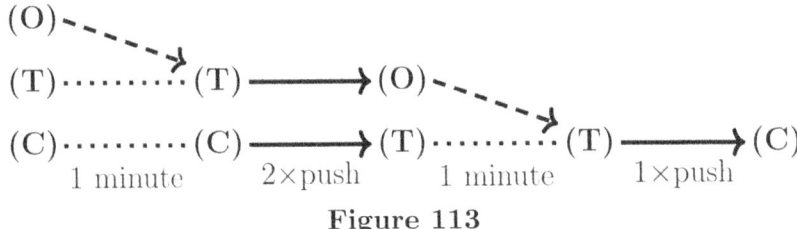

Figure 113

"So the problem can be solved using three pushes of the button," said Tarkal proudly.

"And why is it impossible with fewer pushes?" asked Zsordi.

Tarkal also had an answer to that:

At least one push is needed; otherwise we would be stuck in state (T). If we only push the button once and the lock was in state in (C) initially, then after pushing the button it will be in state (O), and after a minute it will be stuck in state (T). We have not reached state (C).

If we push the button twice and the lock was initially in state (T), it will be in state (C) after the first push and in state (O) after the second push. These transitions are not influenced by a delay between the presses. Subsequently, the system will get stuck in state (T) after a minute, and thus we have not reached state (C).

5.4. Nine Men's Morris Without a Mill

Albrecht assigned numbers to the positions of the pieces according to Fig. 114.

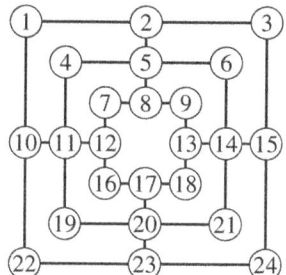

Figure 114

He checked each position in this order. He placed pieces on the first and second positions and skipped the third position to prevent the formation of a mill. He continued assessing the positions based on their numbers: A piece was added at the position with the next number only if no mill was formed; otherwise he left that position empty.

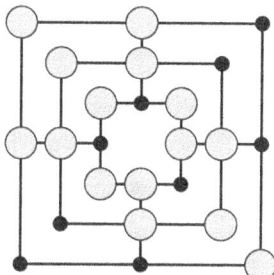

Figure 115

By following this method, he managed to place 15 pieces (see Fig. 115) and contentedly showed his result to Tarkal.

"It is clear that it is not possible to place another piece on the board without forming a mill, so I found the optimal distribution of the pieces."

"I show you another arrangement, have a look at Fig. 116," answered Tarkal. "There are only 12 pieces in this arrangement, however it is still true that if we add another piece to any of the empty positions we will immediately form a mill."

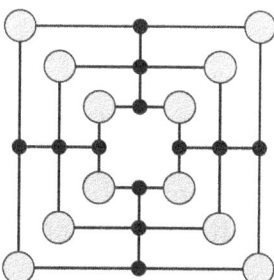

Figure 116

Tarkal's arrangement perfectly illustrates the fact that the inability to place additional pieces without forming a mill does not guarantee that the biggest number of pieces has been placed on the board. Albrecht also understood this.

"So we cannot be sure that 15 is the biggest number of pieces possible. But then how can we prove that a solution contains the biggest number of pieces possible?"

"Let's proceed from line segment to line segment," suggested Tarkal. "It's evident that each line segment can contain at most two pieces."

"However, a single piece is a part of several mills," objected Albrecht.

"Then let's focus only on the horizontal line segments," suggested Zsordi, successfully guiding the team toward the solution below.

 Solution. There are eight horizontal line segments, so they can fit at most $8 \cdot 2 = 16$ tokens. All positions on the board are contained in a horizontal line segment, so it is not possible to place more than 16 pieces.

"I wonder if it's possible to place 16 pieces without forming a mill," wondered Albrecht.

"Surely, if we want to place 16 pieces, then each horizontal line segment has to contain two pieces," said Zsordi. "And naturally, the same must be true for the vertical line segments." It did not take her long to find the following solution with 16 pieces (Fig. 117).

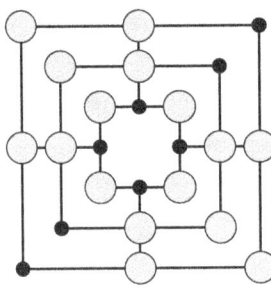

Figure 117

5.5. Length of Line Segments in a Circle

The figure is unusual, and there seems to be no simple method to find the unit length of the square lattice.[16]

 Tarkal's Solution. Let us group the line segments in the figure based on their directions: horizontal, vertical, and skew. In all three cases we will shift the corresponding parts of the diagram to create diameters of the circle.

These shifts are illustrated in separate diagrams, see Figs. 118, 119, and 120. Dashed lines represent the copies of the original segments and arrows to indicate the direction of the shift. In the third figure, two skew line segments

---

[16] The unit length of the square lattice is $1/\sqrt{5} = \sqrt{5}/5$, using the Pythagorean theorem in a similar manner to Problem 4.3.

were shifted, thus forming a radius. The remaining two line segments can be treated similarly, so the skew line segments can be combined to form a diameter.

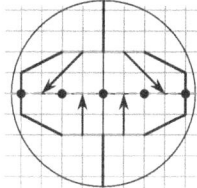
Figure 118

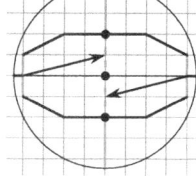
Figure 119

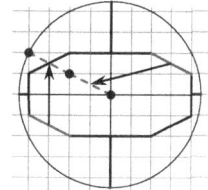
Figure 120

As a result, the total length of the line segments in the original figure is the same as the length of three diameters, i.e., 24 cm.

"Reminds me of the solution to Problem 1.2," said Zsordi.

"It also reminds me of Problem 4.3, skew line segments were also featured there," added Tarkal.

"But there was also a new element in our approach, because the line segments in the figure had to be rearranged," added Albrecht. "The total length of the line segments hasn't changed, but the diagram itself was transformed significantly. I'll create a picture of what happened." (See Fig. 121.)

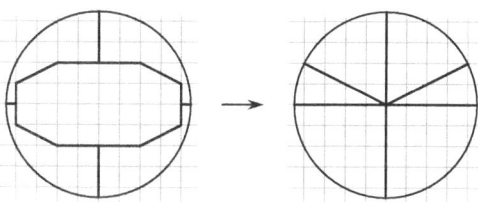
Figure 121

5.6. Jump Over the Other Piece

"How will this game end?" wondered Albrecht and carried on thinking in the following way. "To be able to jump over the piece of the other player, our pieces has to be adjacent. Thus the opponent has to arrive next to my piece with his last move."

"But the other player doesn't want me to be able to jump over his piece, so I can expect him to move next to my piece only if he has no other choice," said Zsordi. "How can I force him to move next to my piece?"

"The way to do this is to have a single empty square between the two pieces," answered Tarkal. "Since the rules do not allow him to jump on my piece, he can only move next to my piece." (See Fig. 122.)

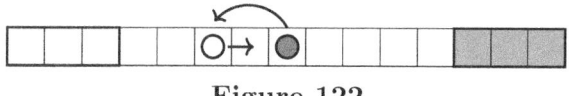
Figure 122

I can make a move resulting in a single empty square between the two pieces if there are two or three empty squares between the two pieces after the other player's move.

If the number of empty squares between the two pieces is 4 after my move, the other player can decrease this number by 1 or 2, and thus the number of empty squares will be 2 or 3.

Before we carry on, let us introduce two new concepts to make it easier to talk about the game. Let us call 1 and 4 *good distance*, because if I have that many empty squares between the two pieces after my move, it will be advantageous for me.

On the other hand, having two or three empty squares between the pieces is a *bad distance*: If this is the distance between the two pieces, my opponent can secure the win. What can we say about bigger distances?

5 and 6 are bad distances, because they can be changed to 4 (i.e., a good distance) in a single step. 7 is a good distance, because it can only be changed to a distance of 5 or 6 in one step.

It is easy to see that the good distances follow each other by a difference of 3: 1, 4, 7, 10, 13, while the remaining possible distances, i.e., 0, 2, 3, 5, 6, 8, 9, 11, 12, 14 are all bad. The winning strategy is to create a good distance in each of my moves. The other player is forced to change it to a bad distance, and I restore the good distance with my next move.

"And what do you do if you can't get a good distance?"

"That's the point! If I've managed to make a good move once, I'll be able to make a good move next time as well. And my opponent will always be forced to make a bad move."

"And if we are at the beginning of the game, and no one has made a move yet?"

"Let's inspect the initial positions of the pieces: the number of empty squares between them can be 9, 10, 11, 12 or 13. Among these 10 and 13 are good distances: in these cases I choose to be the second player. In the remaining cases I choose to be the first player, and with our first move we have to create 7 or 10 (these being the good distances) empty squares between the two pieces."

"That's a good analysis of the game. It reminds me of another game I had read about before our trip."

There is a pile of tokens on the table. Two players take turns playing a game. In each move they can take away one or two tokens from the pile. The player taking the last token wins the game.

"Think about how to play this game. I wonder if you'll also find that it's similar to the previous game," said Zsordi.

"I see the connection," replied Tarkal after some thought. "In fact, I'd say that if we changed the rule so that the player who takes the last token would be the one who loses, then the game would essentially be the same as the game with two pieces."

Finding the reason behind Tarkal's claim is left to the reader.

6. The Trick to Making Moles Disappear

6.1. Blocks of Light Bulbs

Solution. Let us number the rows from top to bottom and the columns from left to right.

The three blocks intersecting the third column contain two, four, and one light bulbs, respectively. The only way to have six light bulbs lit in this column is to have the blocks with two and four light bulbs to be lit, while the block with one light bulb remains unlit (Fig. 123).

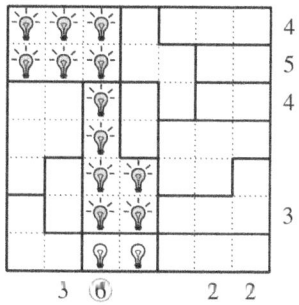

Figure 123

We already know two light bulbs that are lit in the second column. We need one more being lit, and the only way to achieve this is to have the block with one light bulb to be lit and the other blocks to be unlit (Fig. 124).

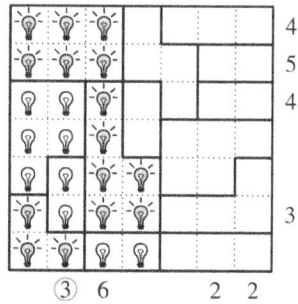

Figure 124

Now consider the first row. Four light bulbs have to be lit, and three are already lit, so the block with one light bulb has to be lit, and the other block has to remain unlit (Fig. 125).

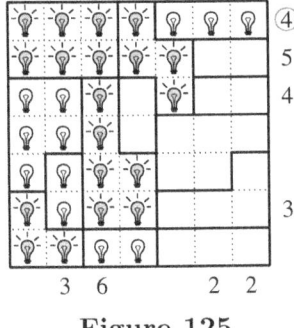

Figure 125

Subsequently there is only one way to complete the second, the third, and the sixth row (Fig. 126).

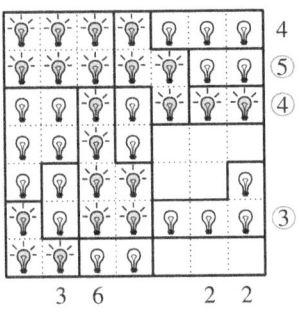

Figure 126

Finally we can finish our solution by inspecting the sixth column. (Fig. 127).

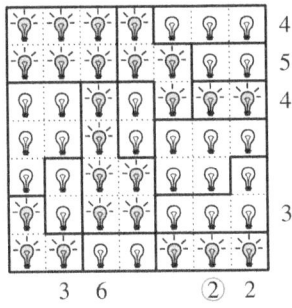

Figure 127

The total number of lit bulbs is 24.

6.2. Rectangle with a Zigzag

 The Solution of Tarkal. Draw perpendiculars to the longer side of the rectangle, forming line segments that create rectangles in the diagram. Furthermore, the zigzagging line divides each of these rectangles into two triangles with equal areas, as the line segments within the zigzagging line are diagonals in

95

the rectangles. Hence, starting from the leftmost triangle, we can determine the area of each small triangle. This method is illustrated in Fig. 128.

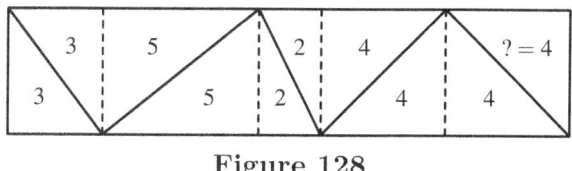

Figure 128

Therefore, the area of the missing triangle is four units.

"I've noticed something," said Zsordi. "The zigzagging line divides the rectangle into two parts with the same area." (See Fig. 129.)

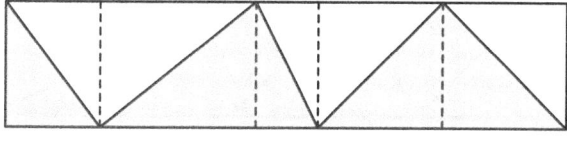

Figure 129

The small triangles at the top and bottom have equal areas, so the areas of the two parts are indeed equal. "Is it true that every zigzagging line divides the rectangle into two parts with equal areas?" wondered Albrecht.

"I believe this can be proved similarly to my method," suggested Tarkal.

"We have skipped an important step," interjected Albrecht. "We have to clarify first what do we mean by a zigzagging line."

"That's true," admitted Tarkal.

"I have a definition," said Zsordi. "A polyline that does not intersect itself, starts from the top left vertex of the rectangle and ends at the bottom right vertex of the rectangle. If the vertices of the polyline are alternating between the bottom and the top side of the rectangle, we call it a *zigzagging line.*"

"I would define a zigzagging line in the same way," agreed Tarkal. "And I believe we can apply my method in general. Let's divide the figure into smaller rectangles using the vertices of the zigzagging line. In each of the small rectangles the area of the top and the bottom triangle will be the same, and this completes our proof."

"This seems to be correct," said Zsordi.

"I'm afraid there is a flaw in the argument," said Albrecht. "I can imagine a zigzagging line where this method does not work. I will quickly draw such a zigzagging line here." (See Fig. 130.)

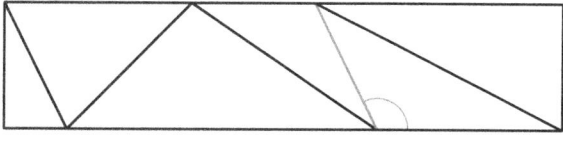

Figure 130

"I have to admit that I haven't thought of that," said Tarkal. "The rectangles created by the vertical line segments can overlap."

"Nevertheless, I still believe that the statement is true. Let's work on it a bit more," suggested Albrecht.

The three friends started to work, and Zsordi quickly came up with the proof.

The Solution of Zsordi. We will give an argument for the zigzagging line that can be seen in the picture. The argument is similar for the general case as well. Let us use the notations in Fig. 131. The dotted lines represent the altitudes drawn from the top vertices of the zigzagging line.

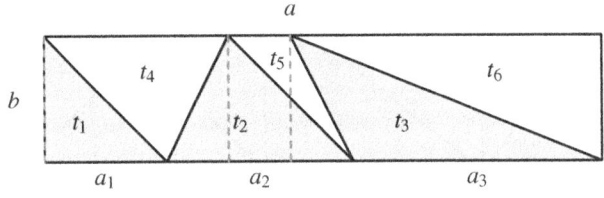

Figure 131

Note that all the heights of the triangles on the bottom are b (the length of the shorter side of the rectangle). We also know that the bases of the triangles on the bottom add up to the length of a (the longer side of the rectangle). Therefore, we can calculate the total area of the triangles on the bottom:

$$t_1 + t_2 + t_3 = \frac{a_1 \cdot b}{2} + \frac{a_2 \cdot b}{2} + \frac{a_3 \cdot b}{2} = \frac{(a_1 + a_2 + a_3) \cdot b}{2} = \frac{a \cdot b}{2}.$$

Consequently, their total area is half of the area of the rectangle.

6.3. Group Stage Qualification

"The only way to score 6 points is to win against two other teams (with a single win the maximum number of points is $3 + 1 + 1 = 5$)," said Albrecht. "If we end up with more points than these teams, then we progressed to the next round."

"Unfortunately, it is still possible to get eliminated with 6 points," replied Tarkal. "If there's a team that lost against all the other teams, it's possible that there is a three-way tie among the remaining three teams. this results in one team having 0 points and three teams having 6 points. If our team has the worst of luck, we will be eliminated. Table 5 shows an example for this scenario."

	csapat	X	Y	Us	Z	W-D-L	Points
1.	Opponent X	–	0-1	1-0	1-0	2-0-1	6
2.	Opponent Y	1-0	–	0-1	1-0	2-0-1	6
3.	**Our team**	0-1	1-0	–	1-0	2-0-1	**6**
4.	Opponent Z	0-1	0-1	0-1	–	0-0-3	0

Table 5

"However, 7 (or more) points guarantee qualification," said Zsordi. "This also means (at least) two wins, and teams that our team defeated can get at most 6 points. These two teams will surely score less points then us."

"So we have established that $A = 6$ and $C = 7$. What about B and D?"

"It's also possible to have a three-way tie with 3 points, see Table 6. The score table is similar to the previous one, just this time team Z won against all the other teams (instead of losing)."

	Team	Z	Us	X	Y	W-D-L	Pont
1.	opponent Z	–	1-0	1-0	1-0	3-0-0	9
2.	**Our team**	0-1	–	1-0	0-1	0-2-1	**3**
3.	Opponent X	0-1	0-1	–	1-0	0-2-1	3
4.	Opponent Y	0-1	1-0	0-1	–	0-2-1	3

Table 6

"However, we still have a chance to qualify with just two points." Check Table 7.

	Team	X	Us	X	Y	W-D-L	Pont
1.	Opponent Z	–	1-0	1-0	1-0	3-0-0	9
2.	**Our team**	0-1	–	0-0	0-0	0-2-1	**2**
3.	Opponent X	0-1	0-0	–	0-0	0-2-1	2
4.	Opponent Y	0-1	0-0	0-0	–	0-2-1	2

Table 7

"However, with only 1 point we definitely won't get to the next round. We have lost at least two times (otherwise we would have at least 2 points). Those who won against us have at least 3 points, so they have scored more points than us, and so we won't progress to the next round."

Therefore, $B = 2$ and $D = 1$. We have completely solved the problem.

When the kids told their solution to the people of Oxisz, they also found it correct.[17] However, 1 week later Albrecht brought up the problem again.

"I feel something is missing. Our answers are probably right, but we weren't thorough enough in our argument. When we established $A = 7$ and $C = 6$,

---

[17]This problem appeared originally in Sixth Dürer Competition. The solution above would have meant a perfect score at the competition. However, one of the organizers found an omission in the argument that will be discussed below.

we had a hidden assumption, namely that there is a dividing line between the points that guarantee getting into the next round and those points which don't."

"This seems to be evident. Why wouldn't it be true? Where did we need it in our argument?"

"We have clearly proved that 7 (or more) points guarantee qualification. We have gave an example where we can be eliminated with 6 points. But how do we know it for sure that we cannot have a score under 6 points which guarantees qualification?" explained Albrecht.

"This sounds really weird," said Zsordi.

"That's what I thought as well," said Albrecht. "But then I've found something interesting. Let's change the rules a bit. Let's have 4 teams in the group, but now a win earns 4 points and draw earns 3 points (losing still earns 0 points). And let the third team also qualify for the next round. Now try to find the number of points that guarantees qualification."

Tarkal and Zsordi thought for some time and then they said: "If every match is a draw, each team will have 9 points. So the least lucky team will be out with 9 points."

"On the other hand, 8 points guarantee qualification. The only way to score 8 points is to win twice and lose once. This guarantees that we qualify, since our opponents only collected 4 points against us, and in the matches between them only $3 \cdot (3+3) = 18$ points will be distributed, so together they will have at most $4 + 18 = 22$ points. So there have to be a team among them who scored less than 8 points, and thus they will be behind us."

"So it's possible to be eliminated with a certain number of points, but surely qualify with a smaller number of points. So there is no dividing line between the scores that guarantee qualification and those with which it's possible to be eliminated."

"At least in the modified game. But what about the original problem, when points are 3/1/0 and two teams progress to the next round?"

"There won't be such a strange phenomenon there. It's possible to show examples when a team won't qualify with 5, 4, 3 and 2 points[18], so C cannot be less than 7. On the other hand, qualification is possible with 3, 4, 5 and 6 points, so D cannot be more than 1. So there is indeed a dividing line between the points that guarantee qualification and those with which it's possible to be eliminated (the dividing line is between 7 and 6 points). Similarly, there is a dividing line between points that guarantee elimination and those with which it is possible to qualify (between 1 and 2 points). But these dividing lines won't exist necessarily and thus they should not be assumed."[19]

---

[18] We will not show the details here, and curious readers are encouraged to find them.

[19] We can use some philosophical terminology to describe the situation: We cannot assume a priori the existence of these dividing lines; however, in many cases a posteriori we can prove that they exist.

6.4. Mole and Bulldozer

"Look! I have an arragement of 10 mole-hills which cannot be bulldozed by Farmer Brown," shouted Tarkal with Fig. 132 in his hands.

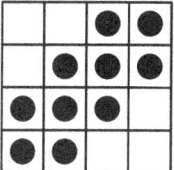

Figure 132

"And why is it not possible to get rid all of them?" asked Albrecht.

"Suppose that Farmer Brown has already completed two columns. The other two columns remain untouched, and the mole-hills in them occupy at least three different rows, regardless of the two columns we have chosen. Farmer Brown can only bulldoze two more rows, which means he won't be able to clear all of the remaining mole hills."

"Two of the moll hills of Tarkal are unnecessary," said Zsordi. "His argument holds true even without them, therefore these eight mole hills are also cannot be bulldozed (Fig. 133)."

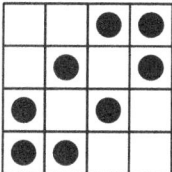

Figure 133

"If I remove any of Zsordi's eight mole hills, Farmer Brown can bulldoze the remaining seven," observed Albrecht. However, the kids have already learned from Problem 5.4 that this does not necessarily mean that Zsordi's solution is optimal. It is possible that there is a better way to arrange a smaller number of mole-hills.

Let us try to look at the problem from the perspective of Farmer Brown.

"How many mole hills can Farmer Brown guarantee to bulldoze?"

"He can surely bulldoze four mole hills in four rounds."

"If there are five mole hills, then there will be at least two in the same column. If Farmer Brown bulldozes such a column first, at most three mole hills will remain. Since he still has three rounds to go, he can get rid of all of them."

"This can be developed further," said Tarkal and presented the following argument.

 Tarkal's Proof That Six Mole-Hills Are Not Enough. Farmer Brown can always bulldoze six mole-hills, regardless of their arrangement.

- If there is a row with three or four mole-hills, let us bulldoze this row first. At most three mole-hills will remain, and therefore in the remaining three rounds Farmer Brown can finish his task easily.

- If every column contains less than three mole-hills, then there must be at least two columns with two mole-hills each (otherwise the number of mole-hills would be at most $2+1+1+1=5$). By bulldozing these two rows only two mole-hills remain, and they can be bulldozed easily in the remaining two rounds.

"Tarkal showed that six or less mole hills are always possible to be bulldozed, and Zsordi showed an arragement with eight mole hills," summarized Albrecht. "The only question remaining is whether we can find a proper arrangement with seven mole hills?"

"If there exists an arrangement with seven mole hills, we have to create three columns, creating two mole-hills each, and the fourth column has to contain a single mole-hill (in other case Farmer Brown can bulldoze two columns with only two mole hills remaining)," mused Zsordi. "Naturally the same can be said about the rows."

We also know that after bulldozing any two columns the remaining mole-hills have to be in at least three different rows.

Taking all these into account, the team quickly found a proper arrangement with seven mole-hills, see Fig. 134.

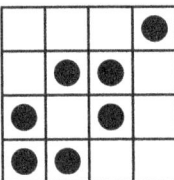

Figure 134

 Checking the Arrangement of Fig. 134. It is enough to check that regardless of the two columns bulldozed by Farmer Brown the remaining two columns contain mole-hills in at least three different rows.

- If any two of the first three columns are not bulldozed, they will contain at least one mole-hill in each of the first three rows.

- And if the rightmost column is not bulldozed, there will be a mole-hill in the top row, and the other untouched column meets at least two of the bottom three rows.

"We have found an arrangement with seven mole-hills that cannot be bulldozed completely, and since Tarkal has already proved that a smaller number of mole-hills can always be cleared, we have finally solved the problem." Albrecht concluded the discussion.

6.5. The Magic Trick of Rodolfo

This magis trick works with many different sequences of suits, but it is very challenging finding a single good sequence. We will refer to the suits with their initials: H, B, L, A.

Zsordi's Sequence. The following sequence of 12 magic cards works:

$$HBLA\ HBLA\ ALBH.$$

What will be the effect of sequence $HBLA\ HBLA$? When H is the magic card for the second time, we can be sure that H emerges on the top of the pile on the table. Indeed, if H is in the hat after the first round, then it will go on the top of the pile after the second time. If H was on the top of the pile after the first round, subsequently B will go on top of it. And B will stay there, because B will not be the magic card before H will be the magic card. So H will surely emerge on the top.

The same argument can be repeated for the other colors. Therefore, after sequence $HBLA\ HBLA$ every card will be on the table in the order of H, B, L, A (from bottom to top).

We know the order of the cards on the table. So sequence $ALBH$ will put all the cards in the hat. Thus we have proved that this sequence works.

"I found a shorter sequence," said Tarkal.

 Tarkal's Sequence. Let the first three cards be HBH. After the first H we can be sure that H is either on the top of the pile or in the hat. The second card makes sure that H is not on the top of the pile. Thus when H is the magic card for the second time, we can be sure that it will emerge on the top of the pile.

Let us continue with sequence LL. Since L is not on the top of the pile, after the first L it will surely emerge on the top. After the second L it will be placed in the hat, and H will be on the top of the pile again.

We can repeat the same process with sequences AA and BB. At this point H is on the table, and the other three cards are in the hat.

Now we pick H to be the magic card, sending it in the hat. Thus we have shown that the following sequence of suits also works: $HBH\ LL\ AA\ BB\ H$.

"What can be the length of the shortest sequence that makes Rodolfo's trick work?" wondered Zsordi.

"I think I can prove that Tarkal's sequence is the shortest possible," said Albrecht.

 Solution of Albrecht. Similarly to problem 5.3, we can show[20] that each suit has to appear at least twice as the magic card.

Every suit has to be the magic suit at least once. If a suit is chosen as the magic suit only once, we cannot be sure that it has ended up inside the hat. Indeed, if it was in the hat initially, it will be placed on the table after being chosen as the magic suit.

---

[20] Although this problem is related to problem 5.3, but it was conceived independently by Kartal Nagy while making the application windows appear and disappear on his computer. The wording of the problem was changed for clarity.

Consequently, every suit has to be the magic suit at least twice. Since we have four suits, the sequence consists of at least eight suits. However, there is still room for a slight improvement of this estimation.

The first suit in the sequence has to be the magic card at least three times. Denote this suit by X. If X is on the top of the pile initially, it will be sent in the hat after the first round, so when X is chosen as the magic card for the second time, X will be placed on the table again. Therefore, X has to appear in the sequence at least three times.

The second type of suit in the sequence also has to be the magic card at least three times. Let the first type of suit be X, and the second type of suit be Y. No matter how many times we pick X, it is still possible that Y is on the top of the pile when it is picked as the magic card for the first time. In this case (similarly to our earlier argument), it is not enough if Y appears in the sequence only twice. So Y also has to appear in the sequence at least three times.

This means that the length of the sequence has to be at least 10. And we have already shown that this can be achieved.

"The number of cards is not important in the problem," remarked Tarkal. "I can show a sequence of length $2n + 2$ that makes the trick work for $n \geq 2$ cards."

"And I can prove that a sequence that is shorter than $2n + 2$ won't work," added Albrecht.

6.6. The Game of Turning Tokens Upside Down

"I suggest considering some easier special cases of the problem first," said Albrecht. "For example, what is a good strategy if there are only one type of token on the table?" This inquiry eventually led to the following solution.

Solution. 1. Blue tokens cannot be turned over, only removed from the table. Therefore, if only blue tokens remain, the problem coincides with the problem mentioned at the end of problem 5.6: *There is a heap of tokens on the table. In each round one or two tokens can be removed from the heap. The winner is the player who removes the last token.* It is easy to see that the winning strategy in this game is ensuring that the number of tokens left on the table after our move is divisible by 3.

2. If only red tokens remain on the table, the second player can win with a very simple strategy: He removes the tokens that were turned on their blue sides by the first player. Consequently, the first player will always find only red tokens on the table, so he has to turn some of them upside down. This way the tokens will run out eventually, securing the win for the second player.

3. If there are only one or two blue tokens on the table (and an arbitrary number of red tokens) and it is our turn, we should take away all the blue

tokens. We have already seen that if only red tokens remain on the table, doing our turn is not advantageous. Therefore, if it is our opponent's turn, we can win by applying the strategy above.

4. However, if there are three blue tokens (and some red ones), we cannot remove all three blue tokens in a single move. There is no point in trying to take away one or two blue tokens, since our opponent could remove the remaining blue tokens and win from there. Turning some red tokens upside down allows our opponent to return to the status quo by removing the newly created blue tokens, thus resetting the number of blue tokens to 3. Eventually the red tokens will be completely removed, and then we will have to remove some blue tokens. Consequently, if there are three blue tokens on the table, we do not want to be the player doing their turn, regardless of the number of red tokens.

5. It immediately follows that having four or five blue tokens (and an arbitrary number of red tokens) on our turn means that we can win by leaving exactly three blue tokens on the table.

6. If the number of blue tokens is 6, an argument similar to the case of three blue tokens demonstrates that we do not want to be the player on turn. And we can carry on with this argument in a similar manner.

It became apparent that only the number of blue tokens decides whether to have our turn. Naturally, this also decides whether it is advantageous to be first or the second player in a given initial position. If the number of blue tokens is divisible by 3, the second player has a winning strategy. Otherwise, the first player can ensure that after his move the number of blue tokens is divisible by 3.

The team sat back contentedly, since they managed to solve the problem.

"Do I understand it correctly that the player following the winning strategy will never touch the red tokens, rather take away some of the blue tokens?" asked Zsordi.

"It's true that he can always win by taking away some blue tokens," answered Tarkal. "However, this does not mean that this is the only way to play this game, there are also other good moves. For example, if there are 8 red and 5 blue tokens on the table, then taking away 2 blue tokens is a good move, but turning 1 red token upside down also works, because this makes the number of blue tokens divisible by 3. If the number of red tokens makes it possible to complete the number of blue tokens into a number divisible by 3, then we can also make a good move by turning some red tokens upside down."

"When I will show this to my little sister, I will use the two kind of good moves alternately," said Albrecht. "It will be harder for her to learn my strategy."

"But if you play against your little sister, it's only fitting that she can decide whether to go first," interjected Zsordi. "And if she chooses correctly, how can you apply the winning strategy?"

"Even if she makes the correct choice initially, a single bad move that leaves the number of blue tokens not divisible by 3 will allow me to decide game's outcome," said Albrecht. "As long as she is not aware of the winning strategy, she'll make random moves, and eventually she'll make a mistake. I have a smart little sister, so eventually she'll discover the good strategy, perhaps with a little guidance from me," Albrecht remarked with a smile.

7. The Rejuvenating Grandma Is Baking a Pie

7.1. Grandma Is Getting Younger

Tarkal's solution. a) Let us observe the age difference between me and my grandma. My age was the quarter of the difference 5 years ago, and now it is the third of the difference. Therefore, the third of the difference is 5 years more than the quarter of the difference.

The difference between the third and the quarter of the age difference equals one-twelfth of the age difference (since $\frac{1}{3} - \frac{1}{4} = \frac{4-3}{12} = \frac{1}{12}$), and we know that this equals 5 years.

Thus the age difference is 60 years. My age is the quarter of this, so I am 20 years old, and my grandma is four times as old, i.e., she is 80 years old.

b) Based on the previous part, we know that my grandma is 60 years older than me. Granny will be three times as old as me when our age difference will be the double of my age. This means that I will be $\frac{60}{2} = 30$ years old, in 10 years' time. My grandma will be $80 + 10$ years old, which is indeed three times my age 10 years from now.

Zsordi's solution. a) Let x denote my current age and y denote my grandma's current age. We know that
$$5 \cdot (x - 5) = y - 5$$
$$4 \cdot x = y.$$

Subtracting the second equation from the first one, we get that $x - 25 = -5$, which yields $x = 20$. Substituting this into the second equation, we get $y = 80$. Therefore, I am 20 years old, and my grandma is 80 years old.

b) Let d denote the number of years until my grandma will be three times as old as me. So we have to solve
$$3 \cdot (20 + d) = 80 + d.$$
Rearranging this gives

$$60 + 3 \cdot d = 80 + d$$
$$2 \cdot d = 20$$
$$d = 10.$$

Therefore, my grandma will be three times as old as me in 10 years' time.

7.2. Too Many Knaves

"What if we consider two cases based on the first person leaving?" asked Tarkal. "If he is a knight, his statement is true, therefore among the remaining 20 people there are at least 11 knaves."

"On the other hand, if he is a knave, then he lied, and among the remaining 20 people there are at most 10 knaves," continued Zsordi.

"This did not bring us closer to the solution. What if we considered another two cases based on whether the second person leaving is a knight or a knave?" suggested Tarkal.

"This would mean 4 cases, and it will only get more complex later," intervened Albrecht. "But I have a better idea. Let's work backwards! Let us consider two cases based on whether the captain is a knight or a knave, and let's focus on the person leaving right before him."

"If the captain is a knight, then the person leaving before him lied, so he must be a knave," began Tarkal.

"On the other hand, if the captain is a knave, the person leaving before him told the truth, so he must be a knight," continued Zsordi.

"Therefore, we can deduce that among the last two people on the ship there's exactly 1 knave and 1 knight," summarised Albrecht.

"Excellent!" rejoiced Zsordi. "Then we can also establish that the person leaving before them was a knave. He said that among them there were more knaves than knights, but this is not true, since there is an equal number of them."

"Indeed!" uttered Tarkal. "And look, we can decide for each subsequent member of the crew whether he is a knight or a knave. I will quickly create a table. I will omit the captain and the member leaving before him, and I will begin with the 19th person, who must be a knave by Zsordi's argument."

"After the 18th person 1 knight and 2 knaves remained on the ship, therefore he told the truth, thus he must be a knight," said Zsordi.

And so they began to fill in the table (see Table 8).

Time of leaving	Remaining knights	Remaining knaves	Thus he is a
19.	1	1	Knave
18.	1	2	Knight
17.	2	2	Knave
16.	2	3	Knight
15.	3	3	Knave
14.	3	4	Knight
⋮	⋮	⋮	⋮

Table 8

"Wow, knights and knaves seem to alternate in the table," observed Zsordi. "Do you think this is just a coincidence?"

Tarkal established "Based on our work so far we can observe that in the odd rows, the number of knights and knaves is the same, while in the even rows, there is one more knave than knight."

"Indeed, and this pattern continues," said Albrecht. "If we examine the next pair, i.e. the 13th and 12th members of the crew, we find that the number of knights and the knaves remaining on the ship will be equal in the next row. Therefore, the next departing member lied, and thus he must be a knave. In the subsequent row we observe that one more knight remained on the ship, thus his statement was true, thus it was uttered by a knight."

"Indeed," admitted Zsordi. "So if we continue filling in the table, the last row would look like this (see Table 9)":

Time of leaving	Remaining knights	Remaining knaves	Thus he is a
⋮	⋮	⋮	⋮
1.	10	10	knave

Table 9

"Wonderful. Therefore, the first person leaving the ship was a knave, and thus there are 11 knaves and 10 knights in the crew of the *Graceful Hippopotamus*," summarized Tarkal.

7.3. Finding the Missing Sums

Albrecht's solution. Arrange the numbers in increasing order. Denote the smallest number by a. Since the sum of the two smallest numbers can only be 105, the second smallest number must be $105 - a$.

The sum of the first and the third number is greater than the sum of the first and the second and less than all the others. For this reason, 111 can only be the sum of the first and third numbers; therefore the third number must be $111 - a$.

Since we cannot decide which two numbers add up to 112, let us focus on 120 instead. 120 can only be the sum of the two largest numbers, and thus the fourth number must be $120 - (111 - a) = 9 + a$.

We have established that the four numbers can be written as a, $105 - a$, $111 - a$, $9 + a$. The sum of two of these must be 112. In addition to the sums already examined, we have three more potential sums, namely $(105 - a) + (111 - a)$, $a + (9 + a)$, and $(105 - a) + (9 + a)$. The second sum is always odd, while the third one is 114, leaving us with $(105 - a) + (111 - a) = 112$. Therefore, $216 - 2a = 112$, and thus $a = 52$.

The four numbers are 52, 53, 59, 61; the six sums are 105, 111, 112, 113, 114, 120. Consequently, the missing sums are 113 and 114.

"I had to work a bit less than you," said Zsordi. "I'll show you my solution."

Solution of Zsordi. Let the four original numbers be $a < b < c < d$. The following can be deduced about the order of their pairwise sums:

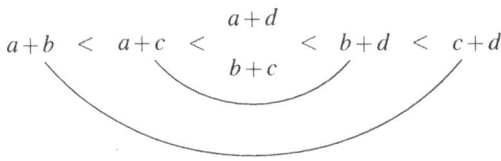

Figure 135

Pair up the sums as shown in Fig. 135. The two numbers in the middle and the two connected pairs all add up to $a + b + c + d$. The sum of the four numbers can also be worked out:

$$a + b + c + d = (a + b) + (c + d) = 105 + 120 = 225.$$

Using the given values, we can find the missing numbers. The fifth sum is $114 \, (= 225 - 111)$, and the fourth sum is $113 \, (= 225 - 112)$.

"Your solution is indeed shorter," said Albrecht. "However, your solution does not reveal the original numbers."

7.4. Dividing the Pie

Solution of Tarkal. I have divided the cake into a 3×3 grid. I only have to divide these nine parts according to the rules. One possible way to do this can be observed in Fig. 136.

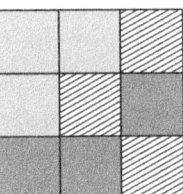

Figure 136

"Something is still missing for me," said Albrecht. "Is it possible to divide the cake in a way that everybody receives a single piece?"

"The question is interesting, and I'm happy to investigate it," said Zsordi. All three kids started to think.

"I have found a solution," exclaimed Zsordi.

 Zsordi's solution. Let us revisit the 3 × 3 grid. Each grandchild should get three portions of the pie and four portions of the crust. We can keep the gray part from the previous solution and adjust the grated and dark parts to get a division satisfying the conditions of the problem, see Fig. 137.

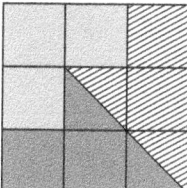

Figure 137

"I've solved the problem in a similar manner, but I've found a slightly more complex division, see Fig. 138," said Tarkal.

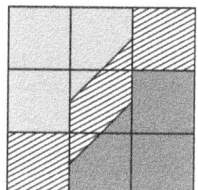

Figure 138

"Indeed, it would be quite tricky to divide a real pie in this manner," added Albrecht.

"I wonder if there is a nicer construction," asked Zsordi.

"What does it mean to be nicer?"

"This a good question," mused Albrecht.

"I suggest requiring the pieces to be convex," said Zsordi.

Is it possible to satisfy Zsordi's requirement? The three friends also began to think about this question.

 Solution of Albrecht. Divide the perimeter of the square into three parts of equal length. Then connect the center of the square with the three points on the perimeter of the square according to Fig. 139.

110

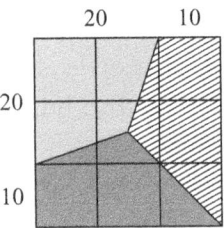

Figure 139

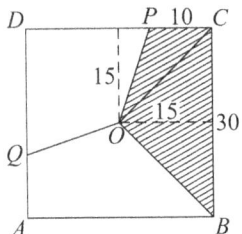

Figure 140

This time we also present an argument that proves that the division is correct, see Fig. 140. The perimeter of each part is 40 cm. Let us calculate the area of the right quadrilateral. Let us divide the quadrilateral into triangles POC and COB.

$$A_{POBC} = A_{POC} + A_{COB} = \frac{10 \cdot 15}{2} + \frac{30 \cdot 15}{2} =$$
$$= \frac{(10+30) \cdot 15}{2} = \frac{40 \cdot 15}{2} = 300 \text{ cm}^2.$$

Since the figure is symmetric, the area of quadrilateral $ABOQ$ is also 300 cm², so the third part has to be of the same area as well.[21] Therefore, this division satisfies the conditions.

"I became hungry... maybe I will bake a pie," daydreamed Albrecht.

"I think Albrecht's solution works for an arbitrary number of grandchildren," mused Zsordi. "While the pie is in the oven, I will try solve the case of seven grandchildren."

Let us allow our main characters baking their pie and thinking, while we delve in this problem in general.

Let us call the division of a shape *fair*, if it satisfies the conditions in the problem, i.e., if the shape is divided into convex parts with equal areas and equal parts of the original perimeter. Is it possible to find a shape different from a square that can be divided among the three grandchildren fairly?

Our method does work for some other shapes, specifically for equilateral triangles, and furthermore, for all the regular polygons. Will it work for other shapes? What property of the square was taken advantage of?

---

[21] We have already seen a similar argument in the solution of Problem 6.2.

The correctness of the construction hinged on the fact that the heights from the common point of the parts were equal. In other words, there was a point that had the same distance from all the sides. Indeed, if we find such a point in a convex polygon and we connect this point to points on the perimeter that divide the perimeter into equal parts, we get a fair division (an example can be seen in Fig. 141).

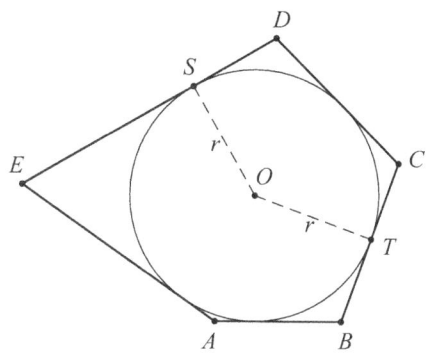

Figure 141

This condition is equivalent to the polygon having an inscribed circle. Filling in the details is left to the reader.

Based on this many convex shapes can be divided fairly. For example, every triangle has an inscribed circle, so every triangle can be divided fairly.

Our method will not work for every convex shape, e.g., rectangles usually do not have an inscribed circle. Is there a method that works for rectangles? This question will be answered in Problem 14.4.

Is there a convex polygon that cannot be divided fairly? This question was answered relatively recently[22] in 1998. The answer turned out to be affirmative: Every convex polygon can be divided fairly. It is no wonder that the solution exceeds the scope of this book. In mathematics, it is common for a seemingly simple question to pose a challenge even for professional mathematicians. Such was the case here.

7.5. Crouching and Standing Students

 a) The solution of Zsordi. Let us imagine the students arranged in a row from the shortest to the tallest and assign them numbers between 0 and 9 corresponding to their heights. Originally I numbered them from 1 to 10, but later I realized that using numbers between 0 and 9 is more advantageous. I also drew pictures of the game for the first few rounds (Fig. 142).

---

[22] Akiyama et al., *Radial perfect partitions of convex sets in the plane*, in: Japanese Conference on Discrete and Computational Geometry. Springer, Berlin, Heidelberg, 1998.

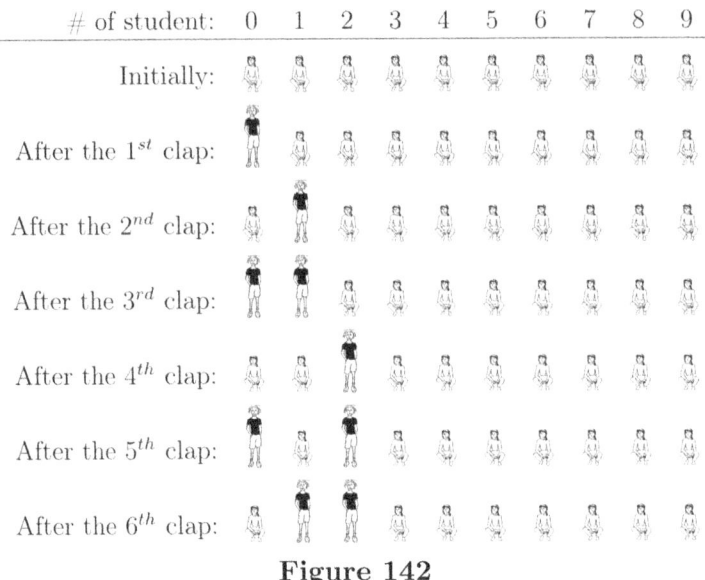

Figure 142

Student #0 (the smallest student) will move after each clap: He stands up and crouches alternately.

What about student #1? When student #0 is standing, student #1 does not move. However, when student #0 crouches, #1 will also move. Thus #1 will move after every second clap, and obviously he will also alternate between standing up and crouching.

Now observe student #2. When he moves, #1 has to move as well: Whenever #2 stands up or crouches, #1 will crouch each time. On the other hand, when #1 crouches, #2 also has to move. We have already established that student #1 will move after every second clap, standing up and crouching alternately, and thus he will crouch after every fourth clap. Consequently, #2 will move after every fourth clap.

A similar argument works for all students: Each will move exactly when the preceding student crouches. Therefore, each subsequent student will move half as frequently as the preceding student.

For example, student #3 will move at every 8th ($8 = 4 \cdot 2 = 2^3$) clap, student #4 will move at every 16th ($16 = 8 \cdot 2 = 2^4$) clap, ..., and the nth student will move at every 2^nth clap.[23]

So student #9 will stand up first after the $2^9 = 512$th clap. At this point all the other students crouch (since when somebody stands up, all the students with a lower number crouch). Therefore, the game essentially restarts for the other students. Student #8 will stand up after another $2^8 = 256$ claps, and the students with a lower number will crouch, the game essentially restarting for them. Student #7 will stand up after another $2^7 = 128$ claps, etc. From here we can see that we need a total of

$$512 + 256 + 128 + 64 + 32 + 16 + 8 + 4 + 2 + 1$$

claps until student #0 finally joins the others standing.

By performing the summation, we find that the answer for the problem is 1023.

---

[23]That is why it is more convenient to start the numbering of the students with 0. If we started the numbering with 1, we would have to write 2^{n-1}.

"Have you noticed that the result, 1023 is exactly $2^{10} - 1$?" asked Kartal after carefully listening to Zsordi's solution. This is not coincidental. Let us imagine an 11th student who is higher than all the other students. We have already established that he would stand up first after the $2^{10} = 1024$th clap. A student will stand up if and only if all the students shorter than him have already been standing. Consequently, the first moment when all ten of the original student are standing will indeed occur only after the $2^{10} - 1 = 1023$rd clap.[24]

 b) Tarkal's solution. Student #8 has not yet stood up (he will stand up first only after the 256th clap), but student #7 has already stood up after the 128th clap, and he is still standing. After the 128th clap, the seven smallest students (#0,#1,...,#6) all crouch, and $200 - 128 = 72$ claps will happen after that. After the $128 + 64 = 192$th clap, student #6 has stood up, so the smaller students have all returned to crouching. After eight more claps, students #4 and #5 have not stood up again. After the last ($128 + 64 + 9 = 200$th) clap, student #3 is standing up, and the smaller students are all return to crouching. Therefore, after the 200th clap, students #7, #6, and #3 are standing (Fig. 143).

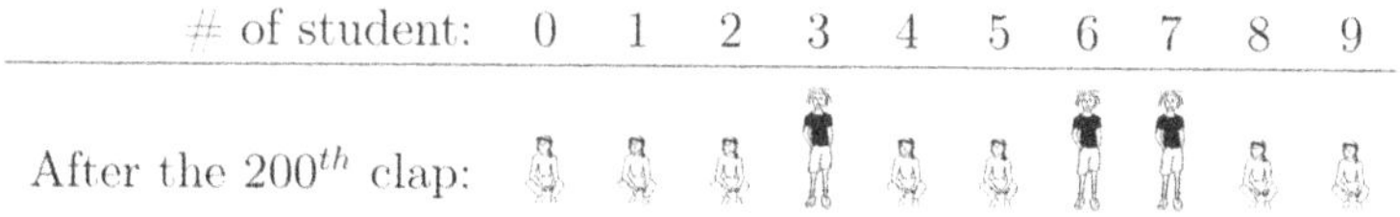

Figure 143

After seeing Tarkal's solution, Albrecht had an epiphany:

"Look! If we reverse the order of the children (so that the smallest child is the rightmost), and substitute standing children with 1's and crouching children with 0's, we would get the base 2 representation of the number of the claps (Fig. 144)."

Figure 144

7.6. The Game of Two Heaps

"How will the game end?" asked Zsordi. "The only way to conclude the game is having two heaps with 1 token each, since neither heap can be divided further. Therefore, the game will be won by the player who can create two heaps with 1 token each."

"Therefore, if we are left with at least one heap with 2 tokens, we remove the other heap from the game, and we win by dividing the heap with 2 tokens."

---

[24]This argument also proves the well known fact about the sum of the powers of 2 between 1 and 2^n being $2^{n+1} - 1$.

"This also means that we shouldn't create a heap with 2 tokens in our turn, because our opponent could win the game immediately."

"However, if we leave a heap with 3 tokens and a heap with 1 token, we will win the game: our opponent will have to divide the heap with 3 tokens, and there are only one way to do that ($3 = 1 + 2$), so he will create a heap with 2 tokens. Similarly, it is also a winning move to leave the opponent two heaps with 3 tokens each."

"This also means that we should not leave the opponent a heap of 4 or 6 tokens, because these heaps can be divided into $1 + 3$ and $3 + 3$ tokens respectively." However, five tokens can be left in a heap, because it can be divided only as $1 + 4$ or $2 + 3$.

"So leaving 1 or 3 tokens in heap is always good, but leaving 2, 4 or 6 is bad," said Tarkal. "I suspect that the parity of numbers play an important role here." Tarkal got it right, and the three friends quickly came up with the following winning strategy.

 Solution. **The winning strategy:** We want to ensure that after our turns there are an odd number of tokens in both heaps. If initially there is a heap with an even number of tokens, we go first and create two heaps with an odd number of tokens. If initially there are two heaps of odd tokens, we let the other player go first.

Proving that the strategy works: On the one hand, since our opponent is left with two odd heaps, he will always create an even and an odd heap on his turn. On the other hand, if at least one of the heaps is even on our turn, we can divide it into two odd heaps. The game can only end when a player creates two heaps of one token. Since these are two odd heaps, following the strategy we will be making the last move.

After the solution Zsordi observed the following.

"In this game there are several ways to make a good move.[25] For example, ten tokens can be divided into $1 + 9$, $3 + 7$, or $5 + 5$ tokens. For example, we could always divide the even heap into $1 + (2k - 1)$ tokens, but our opponent will quickly uncover our strategy. However, if we keep dividing the even heaps randomly, it will be more difficult for our opponent to find the pattern in our moves."

The three friends also made an interesting discovery, which would prove useful in later problems. The games from the previous two sets of problems (Problems 5.6 and 6.6) and this game have the following in common: The winning strategy is based on dividing the positions of the game in two groups: positions to reach and positions to avoid. We suggest calling the former ones *strong positions* and the latter ones *weak positions*. The strong positions of the three games are:

- The odd–odd pairs of heaps in the game of two heaps
- Those positions, where the number of empty spaces between the two tokens gives remainder one mod 3 in the game of the moving tokens (5.6)

---
[25]We experienced something similar in problem 6.6.

- Those positions, where the number of blue tokens is divisible by 3 in the game of flipping tokens (6.6)

Let us examine what does it mean for a position to be strong or weak. The goal is that after reaching a strong position our opponent can only reach a weak position, and from a weak position it is always possible to reach a strong position. The winning position has to be strong, too. Therefore we have deduced the following:

1. From every strong position, it is only possible to move to a weak position.

2. From every weak position, it is possible to move to a strong position.

3. The winning position is a strong position.

If it is possible to divide the positions of a game into strong and weak positions, it is easy to find a winning strategy: In each turn we have to move to a strong position. The first player can win the game if the initial position is a weak position, and one has to allow his opponent to make the first move if the initial position is a strong position.[26]

"And what's up with the first four games we have encountered in Oxisz (1.6., 2.6., 3.6., 4.6)?" asked Tarkal. "What are the strong and the weak positions there?"

"Those are different kind of games. It's only worth finding the strong and the weak positions of a game if the same conditions apply to both players: the winning position and the potential positions are the same for both of them (in other words, after starting the game it does not matter who made the first move, just the positions that were reached in each move)."

---

[26] For this reason many call our weak positions *winning positions*. However, others call strong positions winning positions, because these are the positions reached by the winning player. To avoid this confusion, we introduced adjectives strong and weak to describe the two kinds of positions.

8. Reconcilable Differences

8.1. Erring Organizers?

"I've heard that the Dürer competition is really good, so Zsófi must be right," suggested Zsordi.

"In order to be sure you would have to know all the competitions in the world, but there are too many competitions," objected Tarkal. "I'm almost sure that there are better competitions."

"Let's examine what can be said about the other organizers based on their statements," suggested Albrecht.

"For example, if Zsófi is right, then Bálint is wrong, since he claimed that Zsófi is wrong," observed Zsordi. "Wow, following the given order I can decide for each organizer whether they are right or wrong."

"Me too," added Tarkal.

Summarizing their results, they obtained Table 10:

	If Zsófi is wrong	If Zsófi is right
Zsófi is	Wrong	Right
Bálint is	Right	Wrong
Gábor is	Right	Wrong
Kartal is	Wrong	Right
Dani is	Wrong	Right
Benedek is	Wrong	Right
Magdi is	Wrong	Right
Juli is	Right	Wrong
Bea is	Right	Wrong
Peter is	Right	Wrong

Table 10

"Thus, we can deduce that regardless of whether the Dürer competition is the best competition in the world or not, 5 organizers were wrong," summarized Albrecht. "Therefore, this is the answer to the question."

"Oh, so to solve the problem it wasn't necessary to decide whether the Dürer competition is the best in the world or not," Tarkal pondered. "I feel like I was tricked."

Zsordi's solution. Let us create a diagram of the organizers and their statements about somebody being wrong, see Fig. 145.

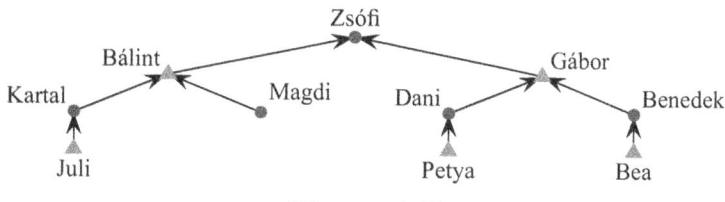

Figure 145

What does it mean that A claims that B is wrong? If eventually A was right, then B would be wrong. However, if A was wrong, then B would be right. In either case, exactly one of them is right, while the other one is wrong.

So if we color people blue and red according to the diagram, we find that either all blue or all red people have to be right.

Since there are five organizers from both colors, exactly five of the organizers will be wrong.

"I feel like I was tricked," Tarkal said dissatisfied. "In neither of the solutions was it possible to decide whether the Dürer competition is the best in the world or not."

"This is not that surprising," answered Albrecht. "You cannot expect the organizers to brag about their competition."

"I agree with Albrecht," said Zsordi. "Moreover, I can tell a new argument based on this."

 Zsordi's meta solution. Based on these statements, the organizers can be divided into two groups: those who consider Dürer to be the best competition in the world (circle) and those who do not (triangle). Since the problem has a unique answer, so the answer has to be the same in both cases.[27] This means that the two groups have to be of the same size, so both have to contain half of the organizers; therefore, exactly five of the organizers will be wrong.

8.2. Find the Connection

"I don't understand how these problems ended up next to each other," said Tarkal upon seeing this problem. "They seem to be unrelated."

"Let's solve them separately, and hopefully the solutions will reveal some connections," suggested Albrecht.

"I will tell a solution to part *a)*," said Zsordi.

 a) Zsordi's solution. I simply arranged the numbers in increasing order.

---

[27] We used this problem in the relay round of the competition. In this round each answer is a number (with at most four digits), which will be checked by the jury, and teams do not have to justify their answers. So every problem must have a unique solution.

```
2255555   2525555   2552555   2555255   2555525   2555552
          5225555   5252555   5255255   5255525   5255552
                    5522555   5525255   5525525   5525552
                              5552255   5552525   5552552
                                        5555225   5555252
                                                  5555522
```

Table 11

I managed to arrange them in a nice triangular shape, see Table 11. In each row the numbers contain the same digits up until the first digit 2, and in each column the numbers contain the same digits starting from the second digit 2. We can easily see that there are $6 + 5 + 4 + 3 + 2 + 1 = 21$ numbers.

Now it was Tarkal's turn with the number of ways to travel from A to B.

 b) Tarkal's solution. For each dot in the picture, I take note of the number of ways it can be reached from A, see Fig. 146.

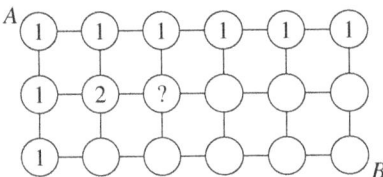

Figure 146

How can I find these numbers? For the dots on the top and the leftmost line, there is just one way to get there, so I filled them out with 1s.

For the remaining dots, I add up the two numbers above and to the left from it. To understand why this works, let us consider point P in Fig. 147.

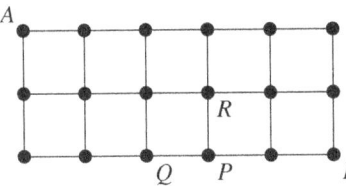

Figure 147

We have been reached point P form point Q or point R. Therefore, every path leading from A to P can be obtained either as a path from A to Q, and then a step to the right, or as a path from A to R, and then a step downward. Therefore, by summing the number of possibilities for each case, we get the number of paths leading from A to P (Fig. 148).

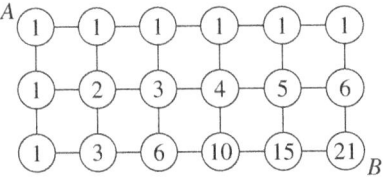

Figure 148

Following this principle we find that there are 21 ways to get from A to B.

Finally, Albrecht discussed the number of points of intersection of the lines.

 c) Albrecht's solution. One line can intersect at most six other lines, so it contains at most six distinct points of intersection. This yields $7 \cdot 6$, but we have to divide by 2, because each point of intersection was counted twice (e.g., the intersection of lines e and f was taken into account on both lines). Therefore, the maximum number of points of intersection is $\frac{7 \times 6}{2} = 21$.

It is also possible to have 21 points of intersection. To reach this number, any two lines should intersect each other (i.e., no two lines can be parallel), and no three lines can meet at the same point.

"The answer to all three problems were 21," said Zsordi. "I suspect this is not a coincidence, rather there is a stronger connection between the solutions."

"I can already see a connection between parts *a)* and *c)*," answered Tarkal. "Here is a new approach for part *c)*."

 c) Tarkal's solution. Two lines can intersect each other in at most one point. If we draw a third line, it can intersect the previous two lines, thus creating two new points of intersection (we have $1 + 2$ points of intersection so far). If we draw a fourth line, it can intersect all three previous lines $(1 + 2 + 3)$. Now it is easy to see that the maximum number of points of intersection for seven lines is $1 + 2 + 3 + 4 + 5 + 6 = 21$.

"Do you remember that the same sum appeared in part *a)*?" asked Tarkal.

"I certainly do, and I can tell the solution of part *a)* in a way that we get the answer as $\frac{7 \times 6}{2}$," answered Albrecht.

 c) Albrecht's solution. There are seven places for the digits.

Choosing the places of the 2s defines the number, because the remaining places have to be filled out with the 5s.

The first 2 can be placed at any of the seven places, leaving six places for the other 2. This would mean 7×6 possibilities. However, we have to divide the result by 2, because we have counted each number twice. For example, number 5525525 can be obtained by placing the first 2 at the third place and the second 2 at the sixth place, but it can also be obtained by putting the first 2 at the sixth place and the second 2 at the third place. So the answer is $\frac{7 \times 6}{2} = 21$.

"What's more," continued Albrecht, "It's also clear that
- if we are looking for the maximum number of the points of intersection of n lines instead of 7,

- or instead of the seven-digit numbers we want to count those n digit numbers that contain only two types of digits, and there are exactly two of one of them,

then we get the following result in both cases:
$$1 + 2 + 3 + \ldots + (n-1) \quad \text{or} \quad \frac{n \times (n-1)}{2}."$$

"But how do we know that both values are the same for every n?" asked Tarkal.

"Exactly because we can solve the problem using either of the two approaches," answered Albrecht.

"But I'll tell you another reason," interjected Zsordi. "Do you remember that in the beginning I have arranged the numbers in a triangular shape? $1 + 2 + 3 + \ldots + (n-1)$ pebbles can also be arranged in a triangular shape. And two such triangles can be combined into an $n \times (n-1)$ rectangle, illustrated in Fig. 149."

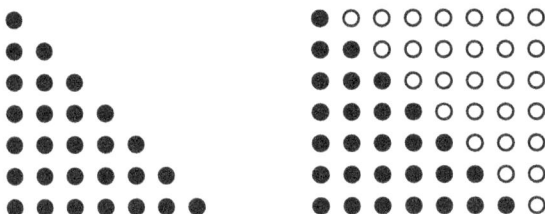

Figure 149

When Albrecht, Tarkal, and Zsordi told their discoveries to a mathematician from Oxisz, he praised them and added the following:

"For expressions $1 + 2 + \ldots + (n-1) = \frac{n(n-1)}{2}$ there is a usual notation[28] of $\binom{n}{2}$. These numbers are also called *triangular numbers* for obvious reasons. The first ten triangular numbers are 1, 3, 6, 10, 15, 21, 28, 36, 45, 55. Since the first triangular number is usually defined as $1 = \binom{2}{2}$, the nth triangular number is $\binom{n+1}{2}$."

We have not yet showed the connection between part *b)* and the other two parts of the problem. Let us check Tarkal's Fig. 150 again.

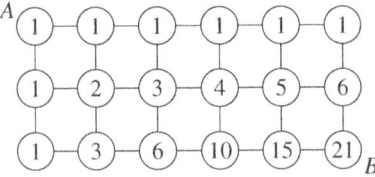

Figure 150

---

[28]The pronunciation of $\binom{n}{2}$ is "n choose 2." In the latter parts of the book, we will use this notation. If the reader is not familiar with this notation, we suggest verifying that $\binom{n}{2}$ answers the following questions: How many ways are there to choose two from n distinct objects, if the order of the two chosen objects does not matter? How many handshakes will n people have, if everybody shakes hands with everybody else?

We can notice that all the triangular numbers appear in the bottom row in increasing order. If we revisit the way we got these numbers from the numbers in the middle row, we will find that 21 was obtained once again as the sum of $1 + 2 + 3 + 4 + 5 + 6$.

But Zsordi found an even more interesting connection between parts *a)* and *b)*.

Connection between parts a) and b) (Zsordi's observation). Consider a path between points A and B (Fig. 151).

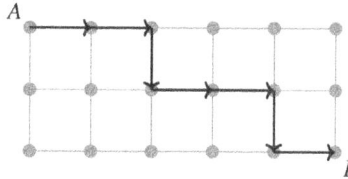

Figure 151

Every such path consists of seven small steps: We have to move to the right five times and downwards two times.

If we know the order of these steps, the path is uniquely determined. For example, if we know that the order of the steps is as in Fig. 152, then we can uniquely determine the path above.

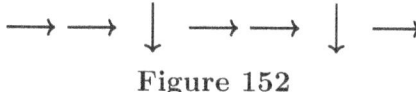

Figure 152

If we consider a sequence of five arrows pointing to the right and two arrows pointing downward and we replace each arrow pointing to the right with 5 and each arrow pointing downwards with 2 (Fig. 153), we get the numbers we had to count in part *a)*.

$$5\ \ 5\ \ 2\ \ 5\ \ 5\ \ 2\ \ 5$$

Figure 153

Therefore, the number of paths in part *b)* is the same as the number of seven-digit numbers in part *a)*.

8.3. Without Adjacent Obtuse Angles

Albrecht's solution. Since the polygon is convex, the measure of its non-obtuse angles is at most 90°. We will refer to these angles as *small* angles.

We will show that a polygon cannot have too many small angles. The external angle[29] of a small angle is at least 90°. So if the polygon has four or more small angles, the sum of the corresponding exterior angles would be at least 360° in total. However, there are also the exterior angles of the obtuse

---

[29]The external angle is the supplementary angle of the corresponding internal angle.

angles, but the sum of the exterior angles of a polygon must be exactly 360°. Therefore, a convex polygon (with at least one obtuse angle) can have at most three small angles.

Consequently, the number of sides can be at most 6 if we want to avoid having two adjacent obtuse angles. A possible example is shown in Fig. 154.

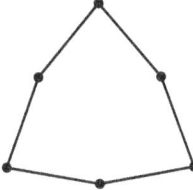

Figure 154

"How do we know that the sum of the exterior angles of a polygon is 360°?" asked Tarkal.

"If we know the sum of the interior angles, it is easy to calculate the sum of the exterior angles," answered Albrecht. "The sum of the corresponding interior and exterior angles is 180°, so the total sum of all the interior and exterior angles in a polygon is $n \times 180°$. The sum of the interior angles is $(n-2) \cdot 180°$, therefore the sum of the exterior angles is $2 \times 180°$."

"I have another argument," added Zsordi. "Let's imagine that we travel around the sides of the polygon with a car starting from the marked point." (See Fig. 155.)

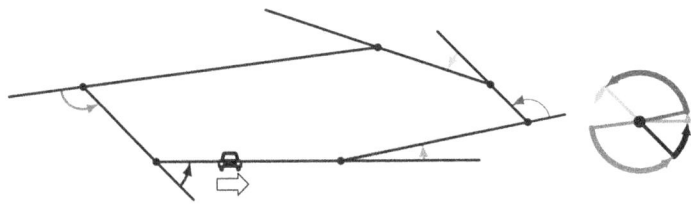

Figure 155

The exterior angles are the measurements of the turns passing from one side to another.

How much do we turn in total? On the one hand, we can track our turns vertex by vertex, accounting for the exterior angle associated with each vertex. On the other hand, upon returning to our starting point, we have completed a full circle, thus making a total turn of 360°. And the total turning equals the sum of the exterior angles, as depicted in the picture. Hence, it follows that the sum of the exterior angles indeed amounts to 360°.

8.4. Quadrilaterals That Are Small and Big at the Same Time

a) "A square cannot be good, because if its side is longer than 10 cm, then its area is greater than 100 cm²," started Zsordi the discussion of the problem.

"What can be said about rectangles?"

"If the length of one pair of sides in a rectangle is longer than 10 cm, the other pair has to be shorter than 0.1 cm (i.e. 1 mm) to get a product (the rectangle's area) smaller than 1 cm². We can imagine a rectangle with small area and two long sides, for example see one in Fig. 156."

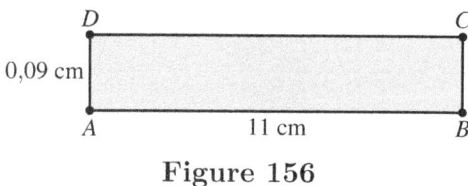

Figure 156

The area of this rectangle is 11 cm · 0.09 cm = 0.99 cm² < 1 cm². Is it possible to increase the short sides while preserving the area?

 a) Solution of Albrecht. If I slide side DC parallel to side AB, the rectangle will become a parallelogram, see Fig. 157. The area will not change, since the area we lose (triangle ADD') is congruent to the area we gain (triangle BCC').

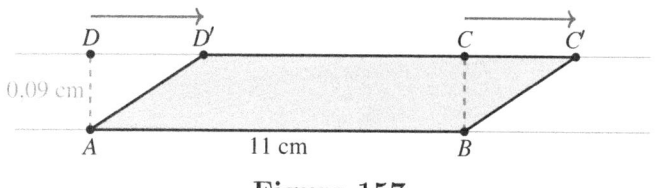

Figure 157

The short sides AD' and BC' are increasing during this process. When vertex D' reaches the original position of vertex C (see Fig. 158), we can be sure that the short sides became longer than 11 cm, since triangle ABD' is a right triangle, and thus hypotenuse AD' is longer than leg AB.

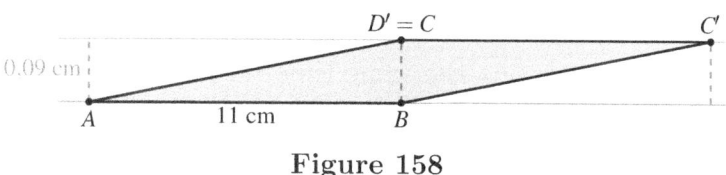

Figure 158

The resulting parallelogram satisfies all the conditions: All of its sides are at least 11 cm long, but its area is at most 0.99 cm².

"I have a different approach," said Tarkal. "I took four straws, each longer than 10 cm, and tried to minimise the area enclosed by them. Since I took for straws with the same length, I ended up getting rhombuses. I noticed that making the rhombuses flatter (making one of their diagonals very short) decreased their areas."

"How did you calculate the areas of the rhombuses?"

"I recalled that the area of a rhombus can be calculated by multiplying the length of their diagonals and dividing the result by 2. But I did not want to calculate the exact lengths of the diagonals, instead I decided to apply some estimations."

a) Solution of Tarkal. If I take a very flat rhombus with sides of length 11 cm, the longer diagonal will always be shorter than the length of two sides combined (because of the triangle inequality), i.e., it will always be shorter than 22 cm. I denoted the length of the short diagonal by x. (See Fig. 159.)

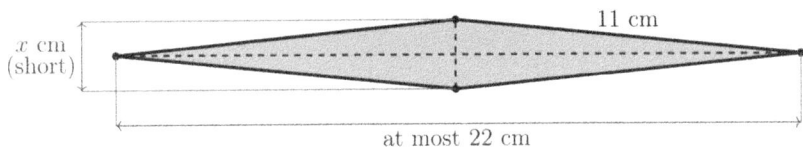

Figure 159

Therefore, the area of the rhombus is less than $\dfrac{x\,\text{cm} \times 22\,\text{cm}}{2} = 11x\,\text{cm}^2$.

If the value of x is chosen such that $11x$ is less than 1, then the area of the rhombus will be less than 1 cm². Is there anything preventing me choosing such an x? It can be arbitrarily small; the worst-case scenario is that the rhombus will be very flat.

The solution of the boys inspired Zsordi to find a third kind of solution.

 a) Zsordi's solution. The quadrilateral could be concave as well, see Fig. 160.

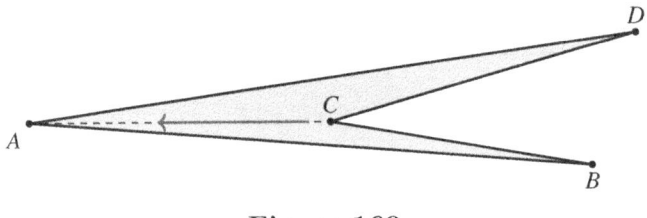

Figure 160

First, I fixed vertices A, B, and D such that sides AB and AD are far longer than 10 cm. If I move vertex C very close to vertex A, the areas of both triangles ACD and ACB will be very small (the sum of these areas equals the area of the quadrilateral). At the same time the lengths of CD and CB will be very close to the lengths of AD and AB, respectively. I have not calculated the exact area and the exact lengths of the sides, but hopefully the essence of the solution is still clear.

Albrecht had the following idea after examining all three solutions:

"It's very interesting that in all of our examples one of the diagonals is very short. Do you think there is a quadrilateral where all four sides and also both diagonals are longer than 10 cm, but its area is smaller than 1 cm²?"

The kids solved this problem later. Although we did not include their solution in the book, we encourage our readers to try solving this problem.

b) "No matter what kind of quadrilateral I draw, if all of its sides are shorter than 1 cm, its area won't be greater than 10 cm². No matter how I look at them, these quadrilaterals are just too small. How can I express mathematically that these quadrilaterals are too small, apart from saying that all their sides are very short?"

b) Albrecht's solution. For example, by saying that, it can be covered with a very small circular disk. Let us draw a circle with a radius of 2 cm around one of the vertices of the quadrilateral, see Fig. 161. Since all the vertices can be reached from this vertex by travelling along at most two sides, all four vertices will be included in the circle.

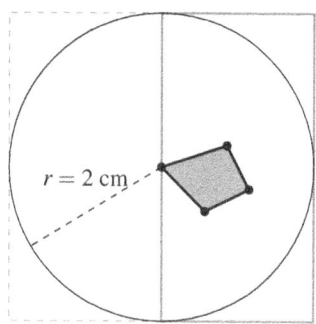

Figure 161

A circle with a radius of 2 cm can be covered with a square of side 4 cm, so its area is less than 16 cm². Although this 16 cm² is still more than the desired 10 cm², our quadrilateral covers only a small part of the circle. For example, if we choose the "leftmost" vertex to be the center of the circle, then it is clear that all the vertices will be in the right semicircle. And the area of this semicircle is less than 8 cm².

"There are several generous steps in your solution, we can see that the quadrilateral takes up only the small part of the half of the square," said Tarkal. "If we delved into the details, we could derive a stronger (i.e. providing a smaller bound) estimate on the area of the quadrilateral. But luckily this is not needed for the solution."

"I will show a sharp upper bound with a completely different method," answered Zsordi.

 b) Solution of Zsordi. Let us divide the quadrilateral into two triangles using one of its diagonals. (For a convex quadrilateral, any of the two diagonals can be chosen, for a concave quadrilateral only the diagonal inside the quadrilateral.)

This way we obtain two triangles with at least two sides shorter than 1 cm.

I will prove that the area of such a triangle is at most 0.5 cm².

Let us label the vertices of such a triangle in a way that sides AB and AC are the short sides. Note that the altitude from vertex C is also shorter than 1 cm.

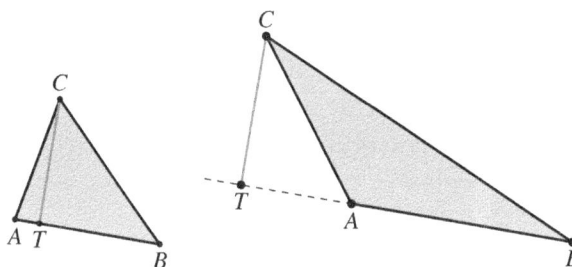

Figure 162

Indeed, if the foot of the altitude is T (Fig. 162), then ATC is a right triangle where the hypotenuse is longer than the legs, so $TC \leq AC < 1$ cm. ($TC = AC$ can occur when T and A coincide, i.e., the angle at vertex A is a right angle.) Applying the area formula of triangles, $\dfrac{AB \times CT}{2} < \dfrac{1 \cdot 1}{2} = 0.5$ cm^2.

The areas of both triangles are less than 0.5 cm^2, so the area of the quadrilateral is less than 1 cm^2. It is easy to see that this upper bound cannot be improved: Let us consider a square where the sides are slightly less than 1 cm.

8.5. The Number of Odd Divisors

"Let's try to find examples," suggested Zsordi.

"I have already found two," said Tarkal. "64 has seven divisors, six even and one odd. 24 also works, because it has two odd (1 and 3), and six even divisors (2, 4, 8, 6, 12, 24)."

"Wonderful. I think 36 also works," added Albrecht. "It has three odd divisors (1, 3, 9), and six even (2, 4, 6, 12, 18, 36)."

"I have observed something," Zsordi exclaimed. "All the answers so far were divisors of 6. I think the solutions of the problem will be the divisors of 6."

"If this is true, there has to be a number that has six odd and six even divisors," suggested Albrecht.

"I've already found such a number," chimed in Tarkal. "90 has six odd (1, 3, 5, 9, 15, 45) and six even divisors (2, 6, 10, 18, 30, 90)."

"It seems my conjecture is correct," said Zsordi. "I will try to prove it."

"I have experimented with other numbers as well," said Albrecht. "I think it is always true that the number of even divisors is divisible by the number of odd divisors."

The kids fell silent. Later, Albrecht found a solution.

 Albrecht's solution. Let us consider an example, $720 = 2^5 \times 3^2 \times 5$. Let us enumerate its odd divisors (1, 3, 5, 9, 15, 45). We have found six odd divisors. We can observe that the doubles of the odd divisors will also be divisors (2, 6, 10, 18, 30, 90).

127

Furthermore, the doubles of these divisors (4, 12, 20, 36, 60, 180) will be also divisors of 720. We can also take the doubles of these and keep on doing this. We can repeat the multiplication by 2 two more times. Thus we have found six odd divisors, and the number of the even divisors is $5 \times 6 = 30$. So in this specific case the number of even numbers is indeed a multiple of the number of odd divisors.

This idea works in general. Let us group the divisors based on the power of 2 in them. We can arrange them in Table 12: The first row contains all the odd divisors. The second row contains all the divisors where the exponent of 2 is 1. The third row contains all the divisors where the exponent of 2 is 2, and so on.

	1	3	5	9	15	45
2^0	1	3	5	9	15	45
2^1	2	6	10	18	30	90
2^2	4	12	20	36	60	180
⋮	⋮	⋮	⋮	⋮	⋮	⋮
2^5	16	48	80	144	240	720

Table 12

Each row contains the same number of divisors, because a new row consists of the doubles of the divisors in the previous row.

It is easy to see that each divisor occurs in the table exactly once. So it is indeed true that the number of even divisors will be a multiple of the number of odd divisors.[30]

Consequently, if the number of even divisors is 6, then the number of odd divisors can be 1, 2, 3, or 6, and we have already showed an example for each.

8.6. Policemen in Cubetown

When the hikers experimented with the game, the policemen easily won the game in part *a)*. Tarkal quickly found a winning strategy.

---

[30]This statement is true for odd numbers as well. The number of even divisors in this case is 0, which is a multiple of every integer.

Tarkal's solution for part a). Move the policemen to the positions in Fig. 163.

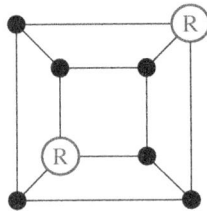

Figure 163

It is easy to see that all the empty corners are adjacent to one of the two policemen. So no matter where the thief moves in the second round, the policemen will be able to catch him (even if the thief makes a double move).

When playing the game from starting position *b)*, it was always possible to move the thief to a corner where the policemen could not catch him with their next move. Furthermore, the double move was not even necessary. Interestingly, if the thief moves twice in a single round, he would find himself in a worse position than before. After his double move, the policemen can catch him quickly.

In the initial position of part *c)*, the policemen can force the thief to make a double move in one of the first two rounds. In order to escape, there is only one way to move in the first round, which can be seen in Fig. 164. Subsequently the policemen can counter by assuming the positions in Fig. 165.

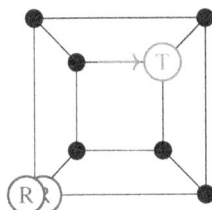 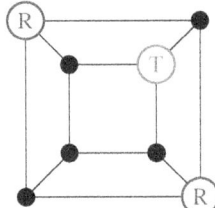

Figure 164   **Figure 165**

In this situation no matter where the thief moves (with a single move), the policemen will capture him immediately.

This may be well, but if the thief uses his double move, the game will become the same as in part *b)*. The thief can always escape from the policemen.

After these observations, Albrecht said the following:

"I'd like to see a proof demonstrating that the thief can escape in part *b)* and *c)*. Additionally, understanding the underlying distinction between the starting positions is crucial. Why do the policemen win in one scenario while the thief wins in the other?"

"The network of streets in Cubetown corresponds to the edge graph of a cube.[31] In Tarkal's solution of part *a)*, the two policemen occupy the opposite vertices of the cube. There are four pairs of opposite corners, and it's true for all of them that if the policemen occupy these two corners, then they will be

---

[31]This explains the name of the town.

able to catch the thief in the next round. But why is it not possible to move the policemen into opposite corners in parts *b)* and *c)*?"

The kids have thought about this for a long time. Finally, the key to the proof was to color the corners with black and white according to Fig. 166.

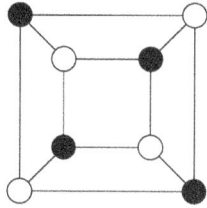

Figure 166

The key property of this coloring is that neighboring corners always have different colors (these are also called *chessboard coloring*). Consequently, in each subsequent round, characters will move to the other color. However, if the thief makes a double move, he will be on the same color as in the previous round.

In part *a)* the policemen occupy corners with a different color, while in parts *b)* and *c)* they occupy corners with the same color. This instantly reveals the underlying reason for the two policemen not being able to reach opposite corners in parts *b)* and *c)*.

"In each round both policemen switches color, thus if they occupied two corners with the same colors initially, then this will stay true for the whole game. Since opposite corners of the cube are of different color, it's impossible for the policemen to occupy two opposite corners at the same moment. However, this does not rule out the possibility that they can win with a different strategy."

However, after a little while the kids completed the proofs.

 Verifying the thief's strategy. b) On the one hand, the thief will never be forced to move to a corner where there is a policeman, because in each round he can choose from three corners (each corner has three neighboring corner). On the other hand, I will prove that the policemen will never be able to move to the thief's corner as long as the thief does not make a double move.

The thief starts from a black corner, so he will move to a white corner in the first round. However, the policemen will move to a black corner from a white corner in the first round, so they will surely avoid the thief.

All three characters switch colors in each round, so in subsequent rounds it will still be true that first the thief can move to a corner not occupied by a policeman with the same color as the corner of the policemen. Now it is the policemen's turn, and they have to move to a corner with the other color. Therefore, the policemen have no chance to move to the thief's corner.

c) All three characters start from a black corner. If the thief uses his double move in the first round, he will still occupy a black corner. Thus the situation familiar from part *b)* will arise, since at the end of each round the policemen and the thief will occupy corners with different colors. Therefore, the thief will win.

"These arguments can be repeated for all initial positions when the policemen occupy corners with the same color," added Zsordi. "It's also easy to see that if the policemen start from corners with different colors, then they can move to opposite corners of the cube even in the first round."

Thus we have solved the game for an arbitrary starting position. If the policemen start from corners with the same color, the thief can win. If the policemen start from corners with different colors, the policemen have a simple winning strategy.

"And it's also easy see that if the thief cannot make a double move, then those initial positions where all three characters start from the same color would also be favorable for the policemen," Tarkal concluded the discussion.

9. How Many Centimeters Is a Dessert?

9.1. Boys, Girls, and T-Shirts

 Zsordi's solution. Let us create a diagram, see Fig. 167.

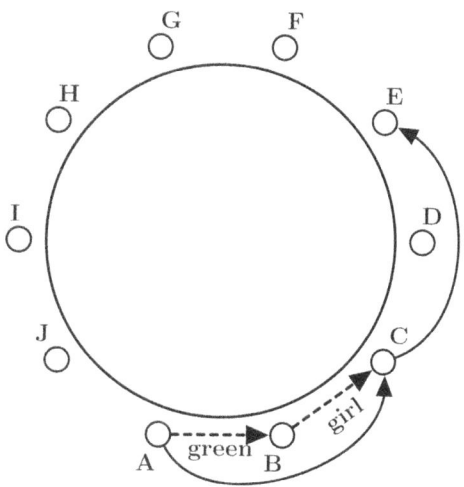

Figure 167

We can assume that A is a girl. The conditions imply that the right neighbor of A wears a green T-shirt; therefore, B wears a green T-shirt. We can make another important observation: C also must be a girl. Indeed, if C was a boy, then B would wear a blue T-shirt (being the left neighbor of C), which is a contradiction.

Thus we have established that the second right neighbor of a girl is also a girl. Therefore, if A is a girl, then C is also a girl, but then also E is a girl. Ultimately we deduce that every second child is a girl.

A similar argument is valid for the boys.

In conclusion, we have derived that every second child shares the same gender. Since both girls and boys are at the table, this situation of feasible only if there are five boys and five girls seated alternately. This scenario is actually possible if all the boys are wearing green T-shirts and all the girls are wearing blue T-shirts.

"What can be said if there are 11 children sitting around the table instead of 10?" asked Albrecht.

"We can still apply the same principle," replied Zsordi. "However, in this case we'll get that all children must have the same gender."

"This problem is similar to Problem 3.5," remarked Tarkal. "On the one hand, in both problems children are sitting around a round table. On the other hand, in both problems we made deductions by checking every second child. In this problem the key insight was deducing that if A is girl, her second neighbour to the right is also a girl. In the other problem, the key was showing that the second neighbour to the right of n is $n+1$."

"In the end, both problems relied heavily on whether the chain of second neighbours form two distinct circles or a single circle containing all the children." added Albrecht.

9.2. Diagonal Halving an Octagon

Albrecht, Zsordi, and Tarkal drew a lot of different octagons. In each octagon they found a main diagonal inside the octagon, so they conjectured that there always exists such a main diagonal. However, they could not prove their conjecture.

"I have found a counterexample," exclaimed Tarkal after a while. See Fig. 168.

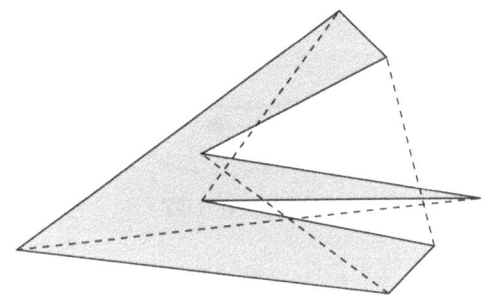

Figure 168

"The answer to this problem is surprising," said Zsordi. "I thought we were on the verge of finding a proof.[32] Is it even true that there always exists a diagonal that is fully inside the polygon?"

"Fortunately, this is indeed true. I remember reading about this statement," replied Albrecht. "However, the proof was rather tricky, and I can't recall it now."[33]

---

[32]We encountered a similar situation when contemplating a challenging problem later featured in the December 2019 issue of *KöMaL*. Our argument rested on the premise that every concave octagon contains an interior main diagonal.
We attempted to prove the statement without any success, and then, much like Tarkal, we have stumbled upon a counterexample. This setback slowed our progress on the original problem; however, as a consolation, a problem suitable for the Dürer Competition was formulated.

[33]The statement and its proof can be found, for example, in Aigner, Martin; Ziegler, Günter (2018). *Proofs from THE BOOK* (6th ed.). Berlin, New York: Springer-Verlag.

9.3. Missing Marks on a Ruler

 The solution of Tarkal. Four marks are sufficient.

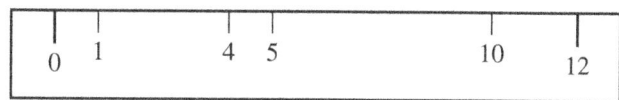

Figure 169

If we add marks at 1, 4, 5, and 10 cm from mark 0 (i.e., at those places where marks 1, 4, 5, and 10 would be on a traditional ruler), as in Fig. 169, we can indeed measure everything as verified in Table 13.

Distance	1	2	3	4	5	6
The two marks	0, 1	10, 12	1, 4	0, 4	0, 5	4, 10
Distance	7	8	9	10	11	12
The two marks	5, 12	4, 12	1, 10	0, 10	1, 12	0, 12

Table 13

Less than four marks will not be enough.

Without a new mark, we can measure a single distance (12 cm). If we add a new mark, we can measure its distance from marks 0 and 12. This means at most two new distances, so at this point we can measure at most $1 + 2 = 3$ different distances. If we add a second new mark, it can be paired up with any of the three previous marks; therefore, we can measure at most $1 + 2 + 3 = 6$ distinct distances. If we add a third mark, it can be paired up with any of the previous four marks, so three new marks will make at most $1 + 2 + 3 + 4 = 10$ distances measurable.

However, we need 12 different distances; therefore, three new marks are not enough, and we have also seen that 0, 1, or 2 marks would make an even smaller number of distances measurable.

"The number of segments created by three new marks can also be calculated in a different way," added Zsordi to Tarkal's solution. "Taking 0 and 12 into account, we have five marks."

And, by using five marks, we can measure at most $\binom{5}{2} = 10$ distances, since each measurable distance requires two marks.[34]

"I really liked your solution, but it has sparked a new question in me," said Albrecht after a little bit of thinking. "With four new marks we have a total of 6 marks on the ruler, so theoretically we could measure $\binom{6}{2} = 15$ distinct distances, which yields a surplus of 3 distances. So if our ruler was 15 cm long instead of 12 cm, would four marks still suffice?"

Zsordi presented the answer next day.

 Zsordi's solution to Albrecht's question. Four new marks will not be enough. Using $2 + 4$ marks, the only way to measure 15 different distances is to make

---

[34] It is also worth recalling the several solutions of Problem 8.2.

sure that all the distances between any two marks are different (since we have $\binom{6}{2} = 15$ pairs to get the same number of distances), and any two marks have an integer distance between 1 and 15 (measured in centimeters). Since 0 and 15 are already marked, all the other marks have to be between them at an integer distance from both. The possible positions of the marks will be denoted by numbers $0, 1, 2, \ldots, 14, 15$, corresponding to their distances from mark 0.

We need two marks at a distance of 14 from each other: This is only possible by placing a mark at 1 or 14, since only pairs $(1, 15)$ and $(0, 14)$ have a distance of 14. Based on symmetry we can assume that there is mark at 1.

Using marks 0, 1, 15, we can also measure distance 1; therefore, we have to be careful not to choose another two marks at a distance of 1 from each other, because then we would not have a chance to get 15 different distances.

For this reason, we cannot mark 2 and 14 (Fig. 170).

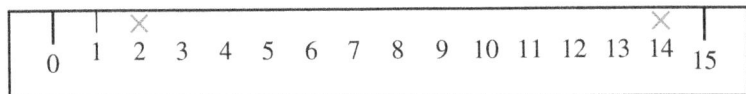

Figure 170

Consequently, we can measure distance 13 only with pair $(0,13)$, so we also have to place a mark at 13.

So far we have four marks at 0, 1, 13, and 15. We can measure distances 1 and 2 with these, and as a consequence 2, 3, 11, 12, and 14 are ruled out, because they are at distance 1 or 2 from an existing mark (Fig. 171).

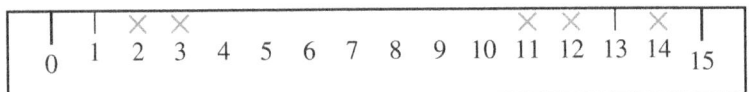

Figure 171

No we can measure distance 12 with pair $(1,13)$, but we also need to measure 11. Based on the numbers already ruled out, it can only be measured by using pair $(4,15)$; thus 4 has to be marked.

Using pair $(1, 4)$, we can measure the distance of 3, and with the pair $(0, 4)$ we can measure the distance of 4. However, this means that a fourth mark cannot be added, because all the remaining marks have distance at most 4 from mark 4 or 13 (Fig. 172).

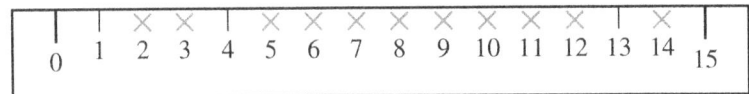

Figure 172

Tarkal also noted—for the sake of completeness—that by using five new marks it is easy to solve the problem, for example, by using marks 1, 4, 7, 10, and 13 (Fig. 173).

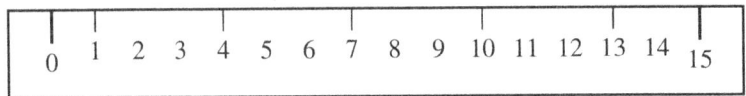

Figure 173

9.4. The Diet of Arthur Noodle

Solution. Let $a \leq b \leq c \leq d \leq e$ denote the number of calories in the individual desserts in increasing order.

The smallest sum is $a+b$. Similarly, the biggest sum is $d+e$. So $a+b = 280$ and $d+e = 590$.

We can also determine the second smallest sum. This will always be the sum of the smallest and third smallest values: $a+c = 300$. Similarly, $c+e = 540$.

Following this path we get stuck, because the next sum cannot be determined uniquely. $b+c = 350$ and $a+d = 350$ are both possible. Now we need a different approach.

All the desserts can be paired up with four other desserts. Therefore, if we add up the ten numbers calculated by Arthur, we get four times the total number of calories:

$$4(a+b+c+d+e) = 280 + 300 + 350 + \ldots + 540 + 590 = 4320.$$

Thus $a+b+c+d+e = \frac{4320}{4} = 1080$.

Notice that now we can determine the middle value. We know that $a+b = 280$ and $d+e = 590$. If we subtract these two sums from the total number of calories (1080), we obtain the middle value:

$$c = (a+b+c+d+e) - (a+b) - (d+e) = 1080 - 280 - 590 = 210.$$

We know values c and $(a+c)$; therefore we can compute a.

$$a = (a+c) - c = 300 - 210 = 90.$$

Knowing a we can find b: $b = (a+b) - a = 280 - 90 = 190$.

d and e can be determined in a similar way. We leave this to the reader and only include the results here: $d = 260$, $e = 330$.

The values of a, b, c, d, and e can only be $90, 190, 210, 260, 330$.[35] It can be checked that the sums are indeed the desired values.

This problem is a natural extension of Problem 7.3. What is more, the ideas seen there (increasing order of sums, the sum of all the values) are also used in this problem.

There is a minor yet interesting difference. In the previous problems all the numbers were specified to be integers. However, this was not the case here,[36] and it was not necessary in the solution.

It is worth rechecking the solution of Problem 7.3 to decide whether the condition was sufficient there? Have we used the fact that the numbers are integers in the solution? Can we get a different solution if the numbers do not have to be integers?

We will return to these questions. In Problem 14.1 we will examine this problem in general. Based on the solution of that problem we will be able to answer the questions raised here.

---

[35] These are the actual numbers of calories in some typical Hungarian desserts.
[36] Although the number of calories is often described with integers.

9.5. Dissection into Rectangle

After a lengthy period of brainstorming, the team of hikers finally arrived at a promising solution to the problem.

"Let's align the target shape and the original one," suggested Zsordi. "This way we can create a large common part."

Zsordi's solution. The missing part of the rectangle consists of 3 and a half squares, and I will call this shape the *boot*. I will carve it out from the hexagon and place it here (see Figs. 174 and 175).

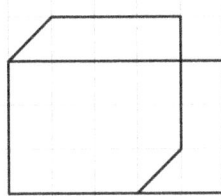 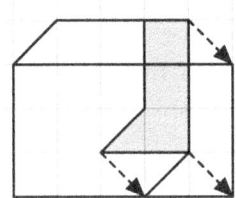

Figure 174   Figure 175

The rectangle is still not complete, but the missing part is not a boot any more, only a *shoe* now. I will also carve out this shoe, and I place it here (see Figs. 176 and 177).

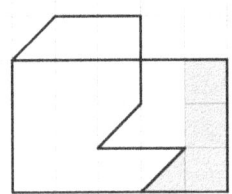 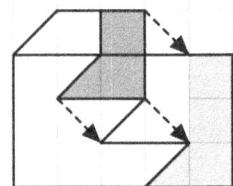

Figure 176   Figure 177

We are almost done. A small part is still missing which I will call the *slipper*. I will carve that out, too, and place it here (see Figs. 178 and 179). Voilà, we are done.

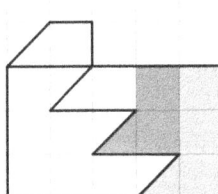 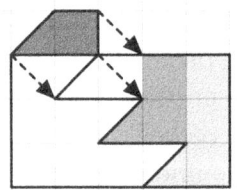

Figure 178   Figure 179

"We're almost done, but we have to be a bit more careful," interrupted Albrecht. "You've divided the hexagon into four pieces, and combined those together to get the rectangle."

137

"Oh, indeed," admitted Zsordi. "But it seemed so pretty."

"Don't worry, it's indeed very nice," said Albrecht. "Luckily, the boot, the shoe and the slipper was moved identically, therefore these pieces can be considered as a single piece, and moved together." (See Figs. 180 and 181.)

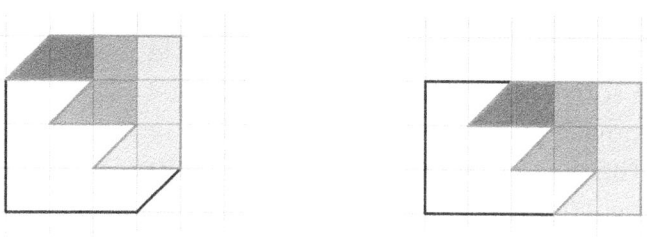

Figure 180   Figure 181

9.6. Double Opening Move Tic-Tac-Toe

The basis of the game is the well known game of *tic-tac-toe*. In the original game, the first player does not have a double opening move; rather both players take turns placing their marks. If a player fills out a row, a column, or a diagonal with his marks (these will be referred to as *triplets*), he immediately wins. If no triplets are formed after filling out all nine squares on the board, the game ends in a tie.

It is well known that if two experienced players play against each other, the game almost always ends in a draw. Both players can follow a strategy that avoids losing the game. A draw is guaranteed with such strategy guarantees, and even winning is possible if the opponent makes a mistake.

These strategies are not based on tricky or beautiful ideas, rather a careful case analysis, since different moves of our opponent require different reactions from us. The strategy for the game of tic-tac-toe is not analyzed in this book.[37]

Albrecht, Tarkal, and Zsordi were already familiar with the game of tic-tac-toe, so they tried to decide which player will gain an advantage from the modified rules. "The double opening move helps the first player, since he can occupy his two favorite (strategically most important) squares on the board at the beginning of the game."

"However, potentially there will also be 5 X's and 4 O's on the board in the game of tic-tac-toe, therefore practically the last X will be chosen earlier. It's not even clear whether this is an advantage."

"Also, the second player is definitely helped by the rule that he will be the winner if both them or neither of them creates a triplet."

"This game is undeniably peculiar. Observe the scenario of Fig. 182. What do you think, which player has an advantage?"

---

[37]Those who are interested can find all the details on the following Wikipedia page: https://en.wikipedia.org/wiki/Tic-tac-toe.

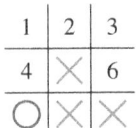

Figure 182

"The first player appears to be in a favourable position. It's O's turn to choose between squares 1 and 2. Therefore, the first player can create a triplet with his next move."

"However, the game does not end here. The first player still has to make sure that the second player does not create a triplet."

"Let's check what happens if the second player places her second O in square 1." (See Fig. 183.)

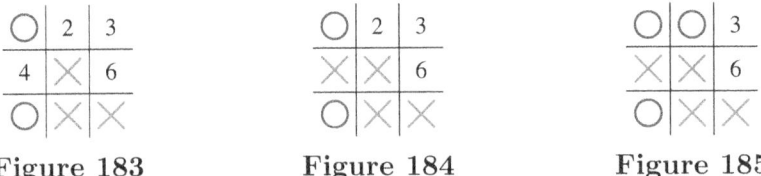

Figure 183 **Figure 184** **Figure 185**

"Now the first player shouldn't put an X in square 2, even though he would create a triplet." "He has to put an X in square 4, otherwise the second player could create a triplet there, and a triplet of O's guarantees a win for him."

"Thus an X will be placed in square 4 (Fig. 184). If the second player places another O at square 2 (Fig. 185), he will win the game. The first player can only decide how to lose the game: if he chooses square 6, both of them will have a triplet, if he chooses square 3, neither of them will have a triplet."

This was surprising, but upon further reflection the children stumbled upon an even greater surprise. From the perspective of the first player, it seems reasonable to put an X in the middle, because this square belongs to four triplets (one row, one column, and two diagonals), while the four corners belong to only three and the four *edges* to only two triplets.

The children repeated the game many times, with the boys consistently placing their first X in the middle, only to be defeated by Zsordi playing as the second player.

"Zsordi, please tell us the reason you always win if we place our first X in the middle," begged Tarkal.

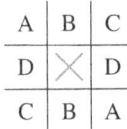

Figure 186

"All right, I will reveal my strategy. The keyword of my approach is 'central symmetry'. I always reflect the last X (not counting the first one in the middle) across the center of the board, and place my O there." (See Fig. 186.)

139

On the one hand, this strategy blocks all the triplets that contain the square in the middle. On the other hand, if a triplet of X's is formed not containing the square in the middle, the mirror images of these three squares will also form a triplet O's.

"However, the first player still has a chance to win the game by not occupying the middle square in the first round," said Tarkal, and he started to try all the remaining openings. After a lot of trial and error, Tarkal finally exclaimed: "I've found a winning strategy for the first player!"

 Tarkal's strategy. Let the first player start with placing two X's in the top two corners as in Fig. 187 (naturally, any two neighboring corners would suffice).

Figure 187

If the second player chooses to place an O between the two X's, then place an X in the center, as in Fig. 188.

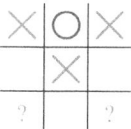

Figure 188

The bottom two corners are still empty. Although the second player can occupy one of the two corners, the first player can always create a triplet in a diagonal by occupying the last empty corner. (If the second player did not choose a corner, the first player can freely choose between the two remaining empty corners.)

Figure 189

Therefore, the first player has a triplet of X's in a diagonal, as in Fig. 189. The second player cannot create a triplet even by marking all the remaining squares on the board; thus the first player wins the game.

However, the second player can choose to avoid placing an O between the two X's. In this case the first player can create a triplet in the first row, see Fig. 190. To guarantee winning, the first player has to make sure that the second player will not create a triplet.

Figure 190

Since we have occupied the first row, no three O's can be formed in a column or a diagonal, since each of these triplets contains a square from the top row. Hence, the first player must ensure that the second player cannot occupy the second or third row, i.e., he has to leave at least one mark in both of these rows. This can be achieved easily in the remaining two rounds; I leave the verification to you.

10. Self-intersecting Tic-Tac-Toe

10.1. Ice Creams on a Stick

a) Albrecht's solution. Let us create pairs from the sticks: For ten pairs we get ten ice cream bars ($\sum : 10$).[38] After eating these we can get five ice cream bars ($\sum : 15$). From the remaining five sticks, we can get two ice cream bars ($\sum : 17$), and one stick remains. Now we have three sticks, so we can get one ($\sum : 18$) and another one ($\sum : 19$) of the ice cream bar. Thus we have obtained 19 ice cream bars.

"Have you also solved the problem with 107 sticks?" asked Zsordi.

"Yes, but first I want to talk a bit more about this method," replied Albrecht. "I call this solution the *halving method*. The method works even easier when the number of sticks is a power of two. For example, if you have 64 sticks instead of 20, the halving method yields $32 + 16 + \ldots + 1 = 63$ ice cream bars, with one stick remaining in the end. Similarly, with 2^k sticks we can get $2^k - 1$ ice cream bars, with one stick remaining in the end."

"How do you count the number of sticks?" asked Tarkal.

"In Problem 7.5 we have already seen that by adding up the powers of two from 1 to 2^{k-1} we get $2^k - 1$," answered Albrecht.

b) Albrecht's solution. Let us create heaps from the sticks with their sizes being different powers of 2 ($107 = 64 + 32 + 8 + 2 + 1$), and apply the halving method for each heap (see Fig. 191).

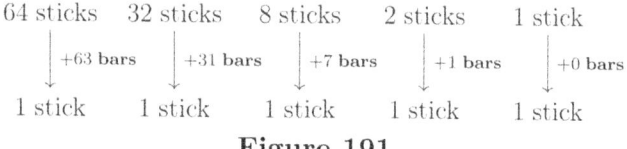

Figure 191

Thus we have obtained $63 + 31 + 7 + 1 + 0 = 102$ ice cream bars, and we are left with 5 sticks. The five remaining sticks can be exchanged in several ways, for example, let us pair up the first stick with the rest. Each pairing will get us

---

[38] This is the notation for the number of bars we got so far.

a new ice cream bar, and we can assume that each time we get back the first stick. Hence, the total number of ice cream bars we got is $102+(5-1)=106$.

"20 sticks yielded 19, while 107 sticks yielded 106 ice cream bars," said Tarkal. "I have a feeling that n sticks will always yield $n-1$ ice cream bars."

"I had the same feeling, but couldn't prove it," said Zsordi. "I tried applying the halving method, but it became very hard to follow."

"Maybe the method of forming a sum of the powers of two can work in general," mused Albrecht.

"That's possible," said Zsordi. "However, we haven't even proved that we cannot get a bigger number of ice cream bars with a different method."

"Oh, I should have thought of that earlier," exclaimed Tarkal. "I've found a very simple solution."

 Tarkal's solution. First note that the number of ice cream bars obtained does not depend on whether we eat the ice cream bar right after the exchange of two sticks. For the next part, we assume that we always do this.

What happens when we exchange two sticks? The number of sticks decreases by 1 (we hand in two sticks and we get back one), and the number of ice cream bars obtained increases by 1. Therefore, the sum of the number of ice cream bars we have obtained and the number of sticks remains constant. The initial value of this sum is n, since we have obtained no bars, and we have n sticks. As long as we have at least two sticks, we can exchange them for another ice cream bar. After exchanging the last pair, we are left with a single stick. Therefore, we have a single stick in the end, and thus the number of ice cream bars we have obtained is $n-1$ (Fig. 192).

Figure 192

"This is indeed much simpler than our methods so far," admitted Albrecht.

"I don't understand why we haven't come up with this solution earlier," said Zsordi, clueless.

What is the reason that the three friends have not come up with the simple solution earlier?

The wording of the problem probably plays an important role. Consider the following scenario: On a hot summer day a large group of people relaxes on the beach. One of them volunteers to exchange sticks to get ice cream bars for the others. In a real-world situation like this, nobody would entertain the curious notion of exchanging the sticks in pairs, which would surely upset the others by taking much longer.

Therefore, discovering the third approach is not straightforward, and one must be able to prescind from the actual situation. This is precisely what makes this problem challenging.

10.2. Enthusiasm and Half-heartedness

(*a*)) The three friends started to work on the problem.

"I've solved it, there are 6 enthusiastic inhabitants," exclaimed Tarkal almost immediately.

"I also think this is the correct answer," said Zsordi. "Have you also checked that there are no other solutions?"

"Well, this is obvious," answered Tarkal. "There are no other solutions, because neither more nor less is possible... I'm afraid I cannot phrase it more rigorously."

Soon Zsordi came up with a full solution.

Zsordi's solution. Let us arrange the answers in increasing order, and place them on a number line, see Fig. 193.

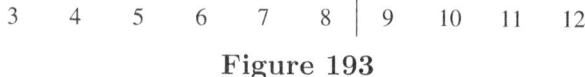

Figure 193

Our initial observation is that no half-hearted person can be to the right of an enthusiastic person. Consequently, the half-hearted answers are on the left part of Fig. 193, while the enthusiastic ones are on the right part of the same diagram.

Let us assume that the line of separation between them is the vertical mark in the diagram. We can deduce that the number of enthusiastic people is at least 8 and at most 9, because the inhabitant saying eight must be half-hearted, and the inhabitant saying nine must be enthusiastic.

On the other hand, it is easy to count the number of enthusiastic inhabitants by counting the numbers to the left from the vertical mark (in this specific case it is 4).

The two conditions on the number of enthusiastic inhabitants have to be consistent with each other. Taking this into account, the vertical mark has to be between 6 and 7. Indeed, if it would be to the right from these, then the number of enthusiastic answers to the right from the mark would be at most 5; however, based on the answers the number of enthusiastic inhabitants should be at least 7. Similarly, if we would move the mark to the left, the number of enthusiastic inhabitants to the right from the mark should be at least 7, but based on the answers the number of enthusiastic inhabitants can be at most 6.

Thus the number of enthusiastic inhabitants can be uniquely determined: There are six enthusiastic members in the group.

(*b*)) After some careful considerations, Albrecht managed to prove that any sequence of integers can be the answers of a group of inhabitants, and the number of enthusiastic members in a group can always be uniquely determined.

Albrecht's solution. Let $A_1, A_2, \ldots, A_{42}$ denote the members of the group, and let $a_1, a_2, \ldots, a_{42}$ denote their answers (Fig. 194).

$$a_1 \quad a_2 \quad a_3 \quad a_4 \quad \cdots \quad a_{40} \quad a_{41} \quad a_{42} \mid$$

Figure 194

Similarly to the solution of the original problem, place a vertical mark separating the enthusiastic and the half-hearted answers. The position of the mark provides two conditions on the number of enthusiastic members.

In the subsequent pictures (Figs. 195, 196, 197, 198, and 199), we will use a square on the number line to illustrate the hypothetical number of the enthusiastic inhabitants. We know that the number of enthusiastic members must be between the two numbers on the two sides of the vertical mark, since according to the half-hearted answers the answer cannot be less than the answer to the left from the mark, and according to the enthusiastic answers the answer cannot be more than the number to the right from the mark. Let us mark this interval on the number line.

If the square is inside this interval, we have found a solution: If the square is at value p, then the number of enthusiastic inhabitants can be p.

We have to prove that there is a unique way of choosing the vertical mark such that the square is inside the interval.

Let us move the vertical mark from the right to the left. Initially, every member is half-hearted, so the square is at 0, the number of the enthusiastic members (Fig. 195). On the other hand, A_{42} is half-hearted, so the number of enthusiastic members must be at least a_{42}.

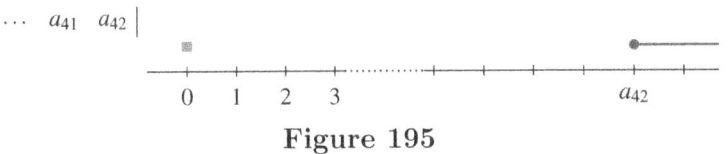

Figure 195

Let us move the vertical mark one place to the left (Fig. 196). Now the square is at 1 (the number of enthusiastic members). On the other hand, A_{41} is half-hearted and A_{42} is enthusiastic, so the number of enthusiastic inhabitants must be between a_{41} and a_{42}.

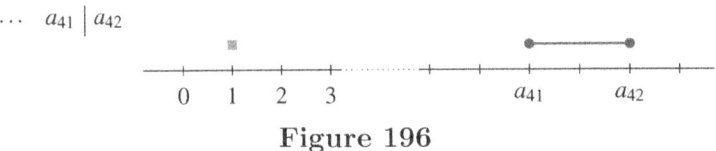

Figure 196

If we keep on moving the vertical mark, the conditions also change. What happens after moving the vertical mark one place to the left? The square will move one place to the right, since the number of answers to the right from the vertical mark increased by one. The interval will move to the left in a way that the left endpoint of the previous interval will become the right endpoint of the current interval.

When the mark reaches the position before the first answer, the interval will contain a_1 and all the numbers to the left from it. In this case, at least some part of the interval will be to the left from the square.

In Fig. 197, the diagram below, we have illustrated the last position of the vertical mark when the square is to the left from the interval.

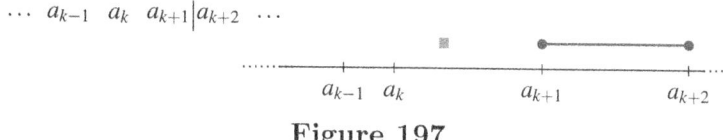

Figure 197

Let us observe what happens in the next step (Fig. 198). The square moves one place to the right, so it cannot jump over a_{k+1}. The right endpoint of the interval will be a_{k+1}, and based on our assumption, the left endpoint of the interval cannot be to the right from the square. Consequently, the square must be in the interval.

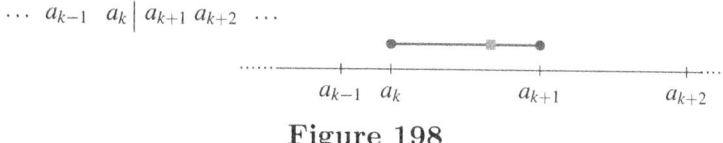

Figure 198

In the next step the square will be to the right from the interval, since it is currently positioned at $(a_k + 2)$, while the right endpoint of the interval is at a_k (Fig. 199).

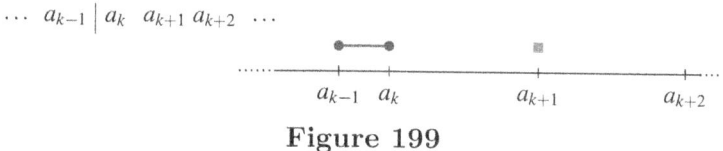

Figure 199

Thus we have proved that for any sequence there is exactly one way to choose the vertical mark in a way that the square is inside the interval. Consequently, the number of enthusiastic members of the group can be uniquely determined for any sequence.

"The second part was pretty tricky," said Zsordi.

"Indeed," said Tarkal, "However, I'm still trying to fully grasp the solution. It seems to me that we haven't used the fact that the answers are integers, though."

"This is not true, we must have used it somewhere," replied Zsordi after some pondering. "No group can provide answers 0, 1/2, 1, 3/2, 2. So our proof cannot be working in this case."

"Nice example. I can already see the part where we had to use that the answers are integers," added Albrecht.

Finding this part is also left to the reader.

10.3. Quadraples of Kritor and Lidek

Solution. It is possible that two different quadruples have the same pairwise sums. Let Kritor choose numbers $2, 3, 5,$ and 8, while Lidek chooses numbers $1, 4, 6,$ and 7. The pairwise sums in both cases will be $5, 7, 8, 10, 11,$ and 13.

"I was pretty convinced that it's not possible to find such quadraples," remarked Zsordi.

"I agree, mainly because in Problem 9.4 we've shown that the pairwise sums of five numbers uniquely determine the original numbers, and we've also provided a method to determine them," added Albrecht.

"I still have some questions," said Tarkal. "How can we find a counterexample? Is there a method or do we have to rely on trial and error? Is it possible to find three quadruples with the same sums? What can be stated about the original numbers?"[39]

10.4. Uniquely Self-intersecting Polygon

a) "Guys, no matter how I try, I cannot draw a uniquely self-intersecting polygon," said Zsordi when the hikers were working on this problem.

"I also couldn't find one," answered Tarkal.

"Then let's try to solve a simpler problem first," suggested Albrecht. "Is it possible to draw a self-intersecting polygon with each of its sides intersecting *at least one* other side?"

"That's not too hard. Here is an example, see Fig. 200," answered Zsordi. "Unfortunately, even three of its sides intersect two other sides."

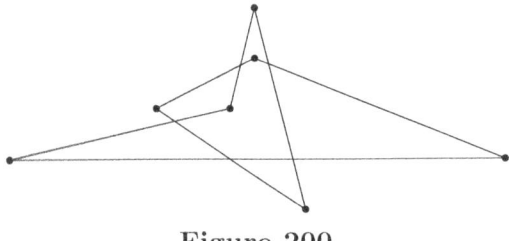

Figure 200

"Is there a way to fix these?" asked Tarkal.

"I have an idea," interjected Zsordi. "Sides that intersect two other sides can be broken up into two sides with a new vertex. So these sides become two sides, each intersecting one other side, see Fig. 201."

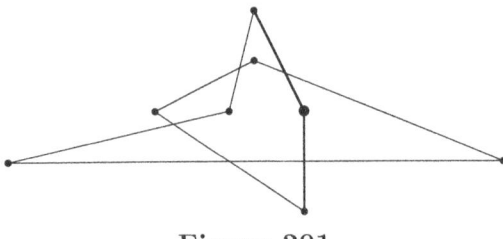

Figure 201

---

[39] In Problem 14.1 we will return to these questions and answer some of them.

"If we repeat this with the other sides intersecting two sides, we get a uniquely self-intersecting polygon, see Fig. 202," Tarkal continued enthusiastically.

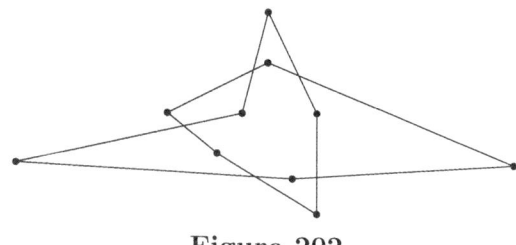

Figure 202

"Indeed, this is a uniquely self-intersecting 10-gon," observed Albrecht.

"Using this method I can also find symmetric examples," continued Zsordi. "In star polygons with an odd number of vertices each side intersects exactly two other sides." (See Figs. 203, 204, and 205.)

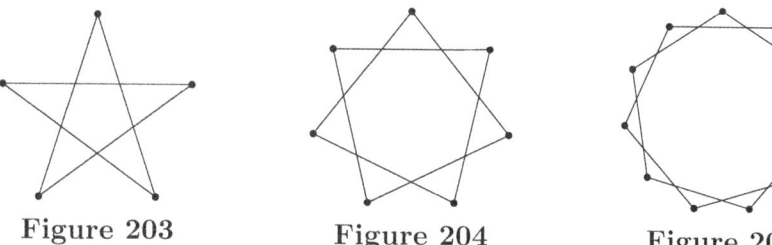

Figure 203 Figure 204 Figure 205

"If we subdivide each of its sides with a new vertex according to our previous method, we get uniquely self-intersecting polygons." (See Figs. 206, 207, and 208.)

b) "We've found several 10-gons," said Zsordi. "Is it possible to draw a uniquely self-intersecting polygon with a smaller number of sides?"

"Star polygons have at least five sides, therefore this way we won't be able to create a polygon with less than 10 sides," answered Tarkal.

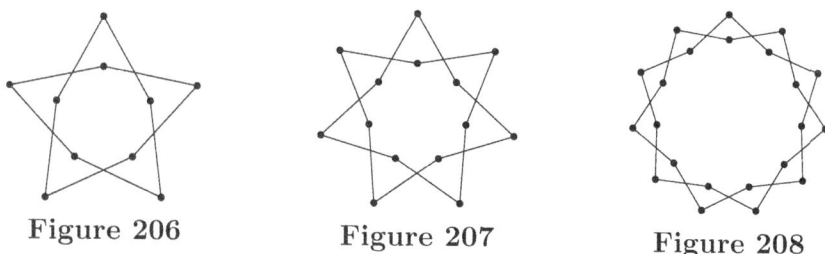

Figure 206 Figure 207 Figure 208

"What if we considered our symmetric diagrams from a different point of view?" suggested Albrecht. "Another way to look at the 10-gon we obtained from the regular star pentagon is to consider two regular pentagons inside each other, and the self-intersecting 10-gon alternates between the outer and the inner vertices, see Fig. 209."

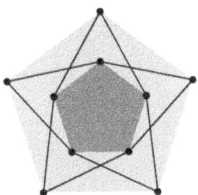

Figure 209

"The same principle can be used to draw a uniquely self-intersecting hexagon built on two triangles, see Fig. 210."

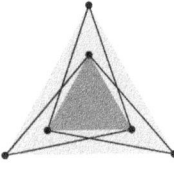 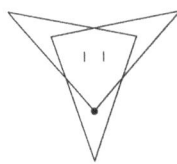

Figure 210   **Figure 211**

"If I turn this self-intersecting hexagon upside down, it will resemble the head of a fox (see Fig. 211), so I will call it a *foxagon*," rejoiced Zsordi.

"So we've found a hexagon, but can we find a solution with an even smaller number of sides?" asked Tarkal.

"I can see that only polygons with an even number of sides can work," answered Zsordi after a little bit of thinking. "The sides of a uniquely self-intersecting polygon can be arranged in pairs, each pair consisting of two sides that intersect each other."

"This leaves us only with quadrilaterals," continued Albrecht. "Let's try to draw such a quadrilateral and call its vertices A, B, C and D. Side AB can only be intersected by side CD, since it shares a common vertex with the other two sides. Thus points C and D must lie on different sides of line AB." (See Fig. 212.)

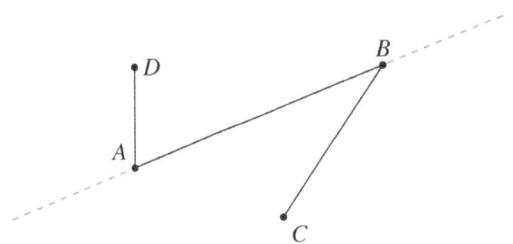

Figure 212

However, this implies that sides DA and BC are also entirely on different sides of line AB, so they cannot intersect each other, and the other two sides also cannot intersect them.

"Thus we have answered part *b)*: the smallest number of sides of a uniquely self-intersecting polygon is 6," Tarkal completed the solution.

"Something is still troubling me," Zsordi said sadly. "I've seen a hexagon and a 10-gon, but is it possible to draw an appropriate octagon? Every attempt to draw one resembling the foxagon results in two separate quadrilaterals (Fig. 213)."

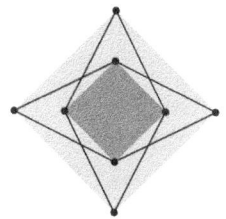

Figure 213

After a few minutes Tarkal slipped a note in front of Zsordi (Fig. 214).

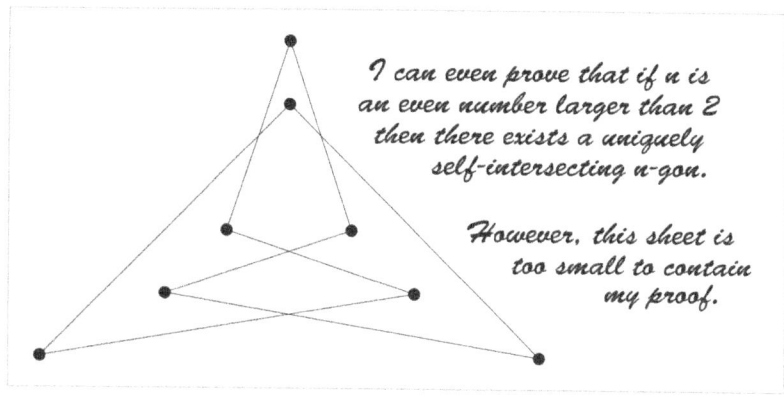

I can even prove that if n is an even number larger than 2 then there exists a uniquely self-intersecting n-gon.

However, this sheet is too small to contain my proof.

Figure 214

10.5. Prime Orders of Divisors

Solution. If a number only has a single prime factor, we can easily solve the problem by arranging its divisors in increasing order.

If a number has two kinds of prime factors, then finding a solution is a bit trickier. Let us use 24 as an example ($24 = 2^3 \times 3$).

Let us arrange its divisors in the way that was described in the solution of Problem 8.5, see Fig. 215.

1	2	4	8
3	6	12	24

Figure 215

One prime order can be 1, 2, 4, 8, 24, 12, 6, 3. Note that after following the first row, the second row is ordered backward.

Now let us consider 120; the prime factorization of which is $120 = 2^3 \times 3 \times 5$.

Among the divisors of 120, we can find the divisors of 24 (1, 2, 4, 8, 24, 12, 6, 3). The remaining divisors of 120 are exactly these divisors multiplied by 5. Let us use the prime order of the divisors of 24, and arrange the divisors in a table again, see Fig. 216.

1	2	4	8	24	12	6	3
5	10	20	40	120	60	30	15

Figure 216

Using the table we can find a prime order: 1, 2, 4, 8, 24, 12, 6, 3, 15, 30, 60, 120, 40, 20, 10, 5.

How to generalize these solutions? Let us consider $300 = 2^2 \times 3 \times 5^2$.

There is a prime order of the divisors that are the powers of 2: 1, 2, 2^2. Let us multiply these with the powers of 3 (only 3 in this case), and arrange them in a table, see Fig. 217.

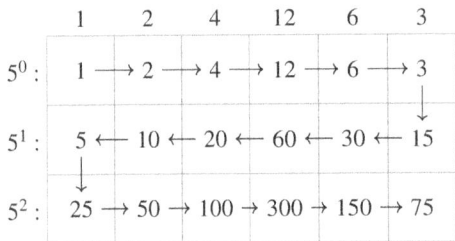

Figure 217

We can find the following sequence by meandering in the table: 1, 2, 4, 12, 6, 3.

Now create a table, see Fig. 218, where the first row contains the divisors of 12, and create subsequent rows by multiplying with the powers of 5.

Figure 218

Order the divisors again by meandering in the table (following the arrows). Why will this be a prime order?

The quotients of the neighboring divisors in each row are primes (because we have chosen a prime order). This is also true for the columns, since the numbers on the top of each other only differ in a single prime factor. Therefore, if we pass from one row to another without changing the column, we get a prime sequence of the divisors.

We can do the same in general. Let us start with enumerating the powers of the first prime factor in increasing order (starting from 1). Let A_1 denote this sequence. Subsequently, let us multiply the divisors in sequence A_1 with the powers of the next prime factor. Let us arrange the divisors thus obtained in a table, and let us order them by following a meandering path, getting sequence A_2 as a result.

We repeat the same method also with the other prime factors. After adding the last prime factor, we get a prime order of all the divisors of our number.

It is worth noting that the sequence always starts with 1; however it does not always end with the number itself.

10.6. Anti-Tic-Tac-Toe

If both players pay close attention, they will avoid choosing a square where a triple[40] is formed with their marks, if possible. Let us see examples of situations where the player on turn is forced to lose the game. In Figs. 219 and 220 it is the first player's (X) turn, and he will lose the game, and in Fig. 221 it is the second player's (O) turn, and he will lose the game.

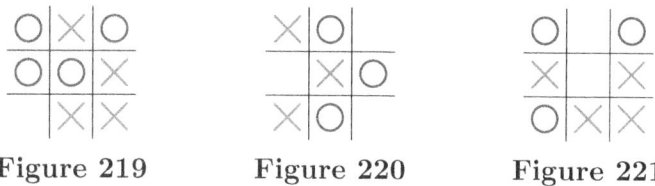

Figure 219   Figure 220   Figure 221

The first player has a disadvantage, since he has to place one more mark on the board, and his last move is fixed, while the second player has a choice in each of his moves.

However, the tie-breaking rule (if no triples can be found after occupying all nine squares, the first player wins) favors the first player.

The team spent a lot of time trying to solve this problem. Ultimately they found the winning strategy by taking advantage of the clever strategy from the version of the game with a double opening move (9.6), where Zsordi found a winning strategy against players starting in the middle square.

 Solution (Winning Strategy for the First Player). Let the first player start with placing an X in the center. (This is a surprising move since the middle square is part of the largest number of possible triples. However, note that the second player can actually force the first player to occupy the middle square by never picking this square.)

Subsequently, in each round place an X at the reflection of the last O *across the center of the board*. This ensures that the middle square will not be in a triple of X's. However, a triple of X's not containing the middle square will not be formed, since the reflection of it across the middle of the board would be a triple of O's that was formed in the previous round by the second player.

However, Albrecht was still very excited to find out how would the answer change if the tie-breaking rule was omitted, and a tie would be allowed. Of course the strategy above guarantees a draw for the first player. But maybe there is an even smarter strategy that will force the second player to create a triple.

The question was ultimately answered by Tarkal, who showed a strategy for the second player to avoid creating a triple.

---

[40]Similarly to the game of Problem 9.6, we will refer to three identical marks in a row, column, or diagonal as a *triple*.

152

Tarkal's strategy to avoid creating a triple as the second player. As the second player we can always place the first two O's essentially as in Fig. 222.

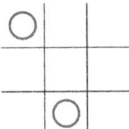

Figure 222

We just have to make sure that we place the first O in a corner such that its opposite corner has both its neighbors free. Since only one X is placed in the first round, this is always possible.

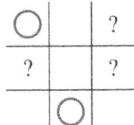

Figure 223

Where should we place the third O? If any of the squares marked with a ? in Fig. 223 are still free, we choose one of those. This way we can guarantee that we will not lose the game: No matter which square with a ? was chosen, there will be at most one square where placing the last O will form a triple. We can choose from two squares in the last round; therefore we can surely avoid such a "dangerous" square.

It is still possible that three squares with a ? are taken by the X's of the first player. In this case the board before the placement of the third O looks like Fig. 224.

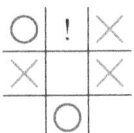

Figure 224

In this case, we place our third O in the square marked with !. At least one of the bottom corners will remain free for the last O, and thus no three of the O's will form a triple.

11. Flowers on the Tiles of the Housing Estates

11.1. Housing Estate

Solution. If only a single house is visible, it must have four floors. And if all four houses are visible, their order from the street must be 1, 2, 3, 4. Based on these, we can create Fig. 225.

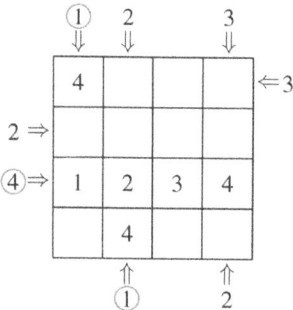

Figure 225

Checking at the second column from the top, the house with four floors must be visible, and the house with two floors is definitely obstructed, because the house with three floors is in front of it. The house with three floors must be visible, therefore we have found the two visible houses, and thus the order of the houses must be 3, 1, 2, 4. (See Fig. 226.)

Checking the first row from the right, only three houses are visible; therefore, the house with one floor has to be behind the house with two floors, and consequently the order must be 4, 3, 1, 2. (See Fig. 227.)

Checking the last column from the top, the order must be 2, 3, 4, 1. (See Fig. 228.)

The remaining two houses in the second row have two and four floors. Since we have already found a house with four floors in the first column, the order in the second row has to be 2, 1, 4, 3. Finally, the order in the last row must be 3, 4, 2, 1. (See Figs. 229 and 230.)

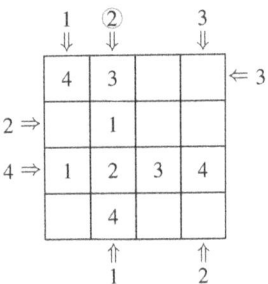

Figure 226

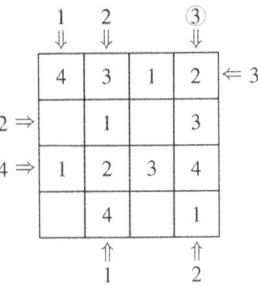

Figure 228

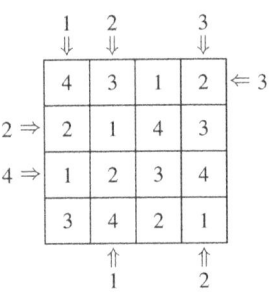

Figure 229

Figure 227

Figure 230

11.2. Bouquets of Flowers

Albrecht solved part *a)* with a clever way of analyzing the cases.

 a) Albrecht's Solution. A bouquet can include:
- Three identical flowers yielding three bouquets
- Three different flowers yielding a single bouquet
- Two flowers of the same type and a third flower of a different type yielding six bouquets. The two identical flowers can be chosen in three ways, and there are two ways to choose the third flower.

Thus there are $3 + 1 + 6 = 10$ ways to create a bouquet of three flowers.

Zsordi applied Albrecht's method for part *b)*.

 b) Zsordi's Solution. We can attempt to analyze the cases again.

Table 14 contains the possible distributions of the types of the five flowers (e.g., $3 + 2 + 0$ means that there will be three flowers from a type, two flowers from another type, and no flowers from the third type):

Distribution	$5+0+0$	$4+1+0$	$3+2+0$	$3+1+1$	$2+2+1$
# of bouquets	3	6	6	3	3

Table 14

For each distribution, the number below them indicates the number of possibilities. If all three summands are different, the number of possibilities is 6: The flower with the biggest number of occurrences can be chosen in three ways, and there are two ways to choose the flower with the second biggest number of occurrences. If there are two equal summands, there are only three possibilities: We have to decide that the flower with the different number of occurrences should be a rose, a tulip, or a gerbera.

Thus the total number of bouquets is $3 + 6 + 6 + 3 + 3 = 21$.

"What do you think, could we answer a similar question with a larger number of flowers, for example 13 or 77?" asked Tarkal.

"I don't feel like applying the method above," answered Zsordi. "I already had to pay close attention not to miss any of the cases. There would be too many cases even for 13 flowers."

"I noticed that for part *b)*, the answer was 21, just like in Problem 8.2," said Albrecht. "And the answer to part *a)*, 10 is also a triangular number. Maybe this is not a coincidence."

"I already see the reason," said Tarkal. "We have consider different cases."

 b) Tarkal's Solution. Let us focus on the number of roses in the bouquet.

- If the number of roses is 5, then there are no other flowers in the bouquet (one possibility).

- If the number of roses is 4, then the remaining flower can be a tulip or a gerbera (two possibilities).

- If the number of roses is 3, then the number of tulips among the remaining flowers can be 0, 1, or 2 (three possibilities).

At this point we can note that if the number of roses is k, the number of tulips can be anything between 0 and $5 - k$. Therefore, the number of possibilities is $(5 - k) + 1$, i.e., $6 - k$. We can keep decreasing the number of roses until we reach 0; the number of bouquets with five flowers is $1 + 2 + \ldots + 6 = \binom{7}{2} = 21$.

"This solution can be easily generalized for any number of flowers."[41]

"In general, from n flowers we can create $1 + 2 + \ldots + (n + 1) = \binom{n+2}{2}$ bouquets, if there are three types of flowers," said Zsordi.

---

[41] We will use the notation introduced in Problem 8.2. For example, from 77 flowers $1 + 2 + \ldots + 78 = \binom{79}{2}$ different bouquets can be created.

"I see an even stronger connection to part *b)* of Problem 8.2," Albrecht added.

 b) Albrecht's Solution. Let us call the horizontal lines of a 5×2 grid Rose street, Tulip street, and Gerbera street (Fig. 231). Consider a path from point A to point B.

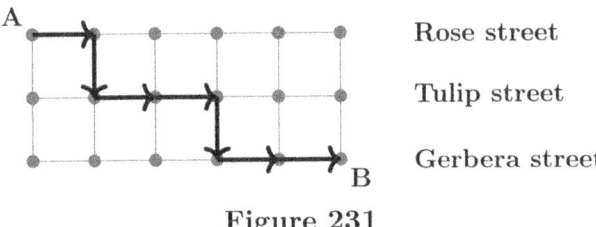

Figure 231

Such a path consists of five steps to the right. All these steps must be in Rose, Tulip, or Gerbera street. For example, the path on the picture makes one move to the right in Rose street and two moves to the right in both Tulip and Gerbera street.

Therefore, this path can be translated into a bouquet of five flowers containing a rose, two tulips, and two gerberas. On the other hand, all bouquets can be translated to a path: The number of roses tells us the number of horizontal steps in Rose street, and then we make one vertical step (if there are no roses in the bouquet, we start with a vertical step). Then I walk along Tulip street for a stretch defined by the number of tulips, make another vertical step, and finally arrive at B along Gerbera street. Thus, there is a one-to-one correspondence between the paths and the bouquets. Consequently, the number of bouquets is the same as the number of paths from A to B.

In general, a bouquet of n flowers from three types of flowers can be translated to a path in an $n \times 2$ grid from the top left to the bottom right corner, where only rightward and downward steps are allowed. This fully explains the number 2s next to and under n in the answer $\binom{n+2}{2}$.

11.3. Tile

"I've found the solution," said Tarkal. "Can I tell it?"

"I think I also have a solution, but I'm happy to listen to you first," said Zsordi.

 Tarkal's Solution. Connect the opposite corners of a standing tile with two "diagonals" that intersect each other in the middle of the tile (see Fig. 232). By doing so, we divide the tile into four pieces. Since the center of both the upper and lower quarter circles is also the center of the tile, the diagonal's length equals the height of a tile; therefore, it is 12 cm long.

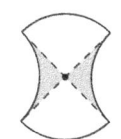

Figure 232   **Figure 233**

If we cut the tile into four pieces with the diagonals (as in Fig. 232), and we place the pieces above, and below the tile (as in Fig. 233), we get two squares. The sides of these squares are half of the diagonals of the tile, so their lengths are 6 cm. The area of the two squares combined together is $2 \cdot 6^2 = 72\,\text{cm}^2$, and hence this is also the area of the tile.

"That was a nice solution," said Zsordi. "I also solved the problem by dissection, but in a different way."

 Zsordi's Solution. The top of a standing tile is a quarter circle, the center of which is the center of the tile, therefore the radius of the quarter circle is 6 cm, and the corners of the tile are also 6 cm from the center of the tile.

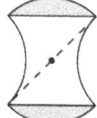   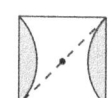

Figure 234   **Figure 235**

Let us dissect the tile by removing the circle segments at the top and the bottom (as in Fig. 234) and placing them at the empty places on the left and right sides of the tile (as in Fig. 235). We got a square with a diagonal of 12 cm. Using the area formula for the kites, the area of the square can be calculated as $\frac{12 \cdot 12}{2} = 72\,\text{cm}^2$.

11.4. The Colorful Flag

When the team of hikers started to work on the problem, Albrecht and Zsordi presented the patterns shown in Figs. 236 and 237 (the numbers in the small squares stand for the colors, and we highlighted colors used multiple times with different shades of gray).

2	1	6	1
1	4	1	8
3	1	7	1
1	5	1	9

3	1	2	7
4	1	2	8
5	1	2	9
6	1	2	10

Figure 236   **Figure 237**

Albrecht's coloring (Fig. 236) is based on assigning the same color to half of the squares in a chessboard pattern. This approach guarantees small squares of the same color in each 2 × 2 part of the flag; therefore the remaining eight small squares can all have different colors. Zsordi (see Fig. 237) created two strips of the same color in the middle of the flag, thus guaranteeing small squares of the same color in each 2 × 2 part, and then assigned different colors to the remaining eight small squares.

Meanwhile, Tarkal settled on dividing the flag into four disjoint 2 × 2 parts, as shown in Fig. 238. We will call these quadrants.

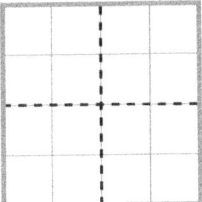

Figure 238

Each quadrant has to contain two small squares with the same color, so each can contain at most three different colors. Therefore, the total number of colors can be at most 12.

The kids decided to work together. They were trying to find colorings with a bigger number of colors, and simultaneously they were trying to improve Tarkal's upper bound.

Solution. Tarkal's bound of 12 colors is not yet sharp. Achieving 12 colors would only be possible if colors are not repeated across different quadrants. However, in this case each of the four squares in the middle (highlighted with a thick frame in Fig. 239) would have a different color, which is not allowed.

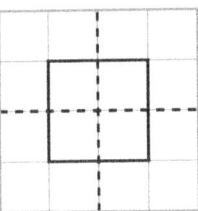

Figure 239

Therefore, we have to relinquish at least one more color; thus, the number of different colors is at most 11.

In addition to the ones we have already analyzed, there are four more 2 × 2 parts where squares with the same colors have to be guaranteed; these are highlighted in Figs. 240 and 241. It seems that may need to sacrifice additional colors, but surprisingly this can be avoided.

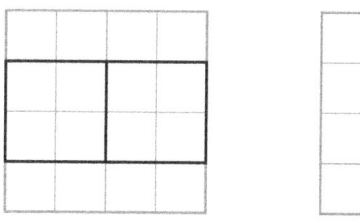

Figure 240 **Figure 241**

If we try using 11 colors in our flag, then we need to use three colors in each quadrant such that there is a single color repeating in two different quadrants, more specifically in the inner 2×2 part of the flag. A quadrant can accommodate three colors only if there is a color appearing twice and two colors appearing once. By choosing the two squares with the same color in smartly in each quadrant, it is possible to handle an additional 2×2 part without sacrificing another color. Such an arrangement can be seen in Fig. 242.

Figure 242 **Figure 243**

This can be completed easily to a flag satisfying the conditions with 11 colors, see Fig. 243.

Later the team also thought about the problem by considering an $n \times n$ square instead of a 4×4 square. This problem turned out to be way harder, nevertheless the exact number of colors can be determined.[42]

11.5. Short Heights and Large Area

If height AT of the triangle ABC is very short, then due to the area formula $\frac{AT \cdot BC}{2}$, the area of the triangle being large means that side BC is very long. Zsordi knew this very well, so she drew triangle ABC with a very long side (BC) and a very short height (AT), see Fig. 244. She also drew the other two heights (BU and CV) and happily realized that they are quite short (at least relative to side BC).

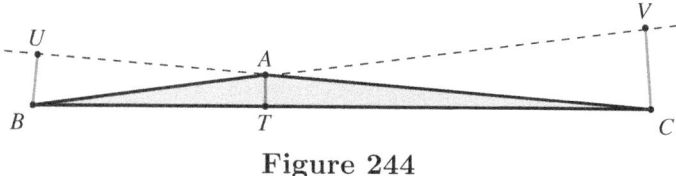

Figure 244

---

[42]This question appeared as problem A.780 of KöMaL in September 2020 (https://www.komal.hu/feladat?a=feladat&f=A780&l=hu).

"I hope that by choosing the length of BC and AT smartly it will turn out to be possible satisfying all the requirements of the problem," said Albrecht. "We can find the area in terms of AT and BC, but it seems to be trickier to keep BU and CV sufficiently short."

"For the sake of simplicity we could choose $AB = AC$ (see Fig. 245), because $BU = CV$ would also be true due to symmetry. If we could find a connection between AT and BU, we would have a good chance at keeping all three heights short simultaneously."

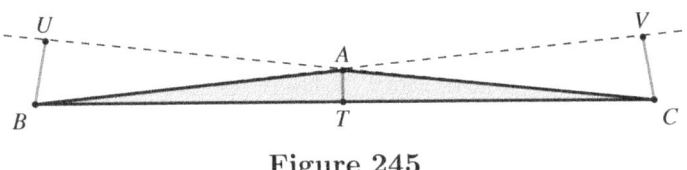

Figure 245

"Have a look at Fig. 246! If we draw rectangle $BCDE$ on side BC which is exactly twice as high as triangle ABC (in other words, A is the centre of this rectangle), then diagonals BD and CE intersect each other at point A. This implies that heights BU and CV are perpendicular to the diagonals."

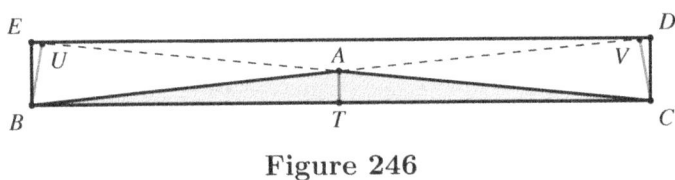

Figure 246

We can also compare the length of BE to the length of BU. Both BE and BU connect point B to line CE. However, the shortest distance between a point and a line is defined by the perpendicular. (We can also argue that BUE is a right triangle with BE being the hypotenuse; therefore, BU is shorter than BE).

"BU turned out to be shorter than BE, which is twice as long as AT. Therefore, if AT has a length of 5 mm, then BU (and also CV) will be shorter than 1 cm."

"We only need to define the length of the base BC. The area of the triangle ABC is $\frac{AB \cdot AT}{2}$. If the length of BC is 1 km, then the area of the triangle ABC is $2 \times 5 \, \text{m}^2$, since the product of 1 km and 1 mm is exactly $1 \, \text{m}^2$."

"This is very interesting. A side with a length of one kilometer would not fit in my notebook, while E and U are so close to each other that I could hardly tell them apart," Zsordi mused. However, she continued, "although it's not possible to actually draw this triangle, we can still imagine it, and we can calculate its dimensions.[43] Therefore, it will provide a solution to the problem."

---

[43] It is worth recalling the solution of Problem 8.4. We already had to choose the lengths of some of the line segments extremely short, so we have not tried to draw a picture to scale. We imagined the exact picture instead of actually drawing it.

11.6. Taking Away Tokens on the Same Line

"We've already seen in both version of the tic-tac-toe (Problems (9.6 and 10.6)) that reflecting the moves of our opponent across the centre of the board can be advantageous. Let's attempt applying it here as well," said Tarkal, and he placed a mark in the middle of the rectangle. "I will denote the centre of the rectangle by K."

 a) Tarkal's Strategy. For the initial arrangement of the 4×4 square, I would like to be the second player. Regardless of the first player's move, I will respond by reflecting his move across K. For example, if the first player removes the tokens highlighted with the dotted line in Fig. 247, I will remove the tokens highlighted with the continuous line in Fig. 248.

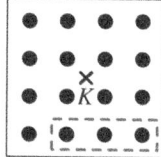 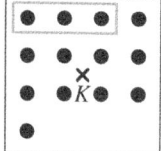

Figure 247   **Figure 248**

Figures 249, 250, 251, and 252 show a possible sequence of moves in a game that was played this way. (The moves of the opponent are highlighted with dotted lines, and my reactions are highlighted with continuous lines.)

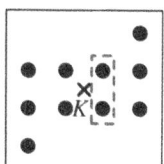 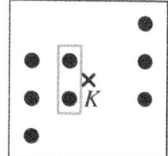 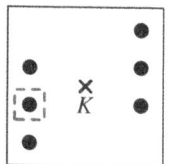 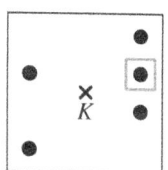

Figure 249   **Figure 250**   **Figure 251**   **Figure 252**

"I can see that you'll win from here. I even tried other moves as the first player, and your strategy seems to be always working," said Albrecht. "I have a feeling that something is still missing, though. You haven't yet proved the reflection of the moves of the first player are always legal. And you also haven't yet proved that you'll always remove the last token."

"I'm also very interested in these proofs. But I'm even more eager to check whether Tarkal'a strategy works for initial positions *b)* and *c)*," said Zsordi and delved into the problem enthusiastically. "In part *b)* it's possible that I won't be able to reflect even the first move of my opponent. The reason being that there is a token in middle which is a reflection of itself. Therefore, if central token is removed, the move cannot be reflected across K. However, this problem can be rectified by removing the central token as soon as possible."

 b) The Strategy of Zsordi. For the arrangement of 3×5 tokens, I would choose to start. My first move is to remove the central token located at point K. Subsequently I apply the strategy of Tarkal, i.e., I always reflect the move of my opponent across C. Figures 253, 254, and 255 show an example for the first three moves.

162

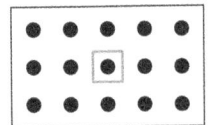 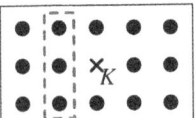 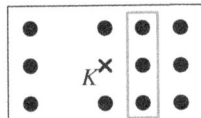

Figure 253   **Figure 254**   **Figure 255**

"I can also see other good initial moves. I could remove the entire row or column in the middle, and still be able to reflect everything with Tarkal's strategy. Additionally, this approach will also help with part *c)*."

"But there is no token on K in part *c)*. Can't we simply apply the reflecting strategy as the second player?"

"It's true that there is no token in the middle of the arrangement of 4×5 tokens, however, there are still moves which cannot be reflected. If the line of the tokens passes through K, it cannot be reflected. We can eliminate this problem again with a smart first move."

 c) The Strategy of Albrecht. Begin by removing the entire middle column as the first player (Fig. 256). Then, it becomes possible to apply the well-tried strategy of reflecting opponent's moves (Figs. 257 and 258).

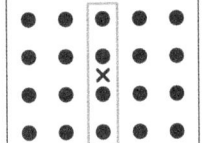 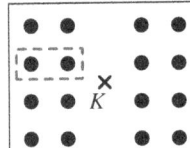 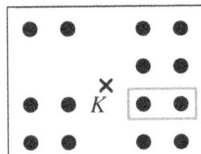

Figure 256   **Figure 257**   **Figure 258**

"As far as I can see, the first move can be any line of tokens that is symmetric with respect to K. Subsequently, reflecting across K becomes possible. In part *a)* there is no set of tokens symmetric with respect to K that can be removed. In part *b)*, all such lines contain the token at K," said Zsordi.

"However, there can be other initial moves that work. For example, in part *c)*, we can also take away the first or the last column in the first round. The remaining tokens form a 4×4 square. It's our opponent's move, and we will be able reflect across the centre of the remaining square," added Tarkal.

"We've learned that the reflecting strategy, with proper preparation, can be applied in all scenarios. It's time for us to prove this."

"What guarantees that reflecting our opponent's move across K will always be a legal move?"

"Let A denote the set of tokens removed by our opponent in his last move. Let B denote the reflection of A across point K (these are the tokens we want to remove). Since the tokens in A are in the same row or the same column, the same holds for B."

"And why can't be any gap in B?"

"Because if B would contain a gap, A would contain its reflection as a gap, therefore it's also not a legal move. After our moves it's guaranteed that the shape formed by the tokens on the table is symmetric with respect to K.

163

Therefore, if B would contain a gap, its reflection across point K, A would also contain the reflection of that gap."

"Wait a minute! And is it not possible that the gap in B was created formed after removing the tokens in A?"

"This is only possible if there is a common token in A and B. Since A is a set of tokens in the same row or column, this is only possible if it passes through K. This means that the token at K would be a part of A, or K would be between two adjacent tokens in A. However, this is not possible, since if there was a middle row or column, we would have removed its central part (the token at K, or the two adjacent tokens on either side of K)."

"This completes the proof. We've proved that we can always reflect our opponent's move (in a legal way), and thus we will have the last say in the game, and we will remove the last token. And it's also clear that our strategy will work for any $k \times n$ rectangle. Based on the parity of k and n we'll get solutions analogous to parts a), b), and c)," summarized Zsordi.

Albrecht still had a question though.

"Do you remember the concept of strong positions introduced in the game of two heaps (7.6)? What could be the strong positions the game of removing tokens?"

"We've essentially proved that those positions are strong where the tokens form a shape with central symmetry, and the tokens in the middle of the middle row or column (should they exist) are removed."

"And could there be other strong positions?"

"Yes, there are, and I will show you three examples (you can check that these are indeed strong positions)." (Figs. 259, 260, and 261.)

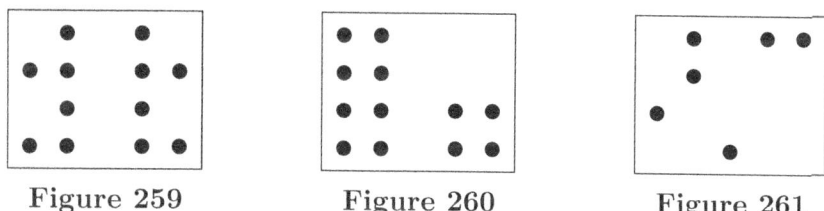

Figure 259   Figure 260   Figure 261

This also illustrates that in some positions it is possible to make a good move that is not a reflection of the previous move.

12. Trolleybus on the Oktogon

12.1. Trolleybus, Bus, Tram

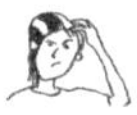

Solution. Let us use a Venn diagram to illustrate the conditions of the problem. Each set contains cities with a given mode of transportation, and we will illustrate the conditions separately (from Fig. 262, 263, 264, and 265).

Light shading indicates parts that contain at least one city, while dark shading indicates parts not containing any cities. Condition *c)* implies that at least one of the two highlighted parts cannot contain any cities.

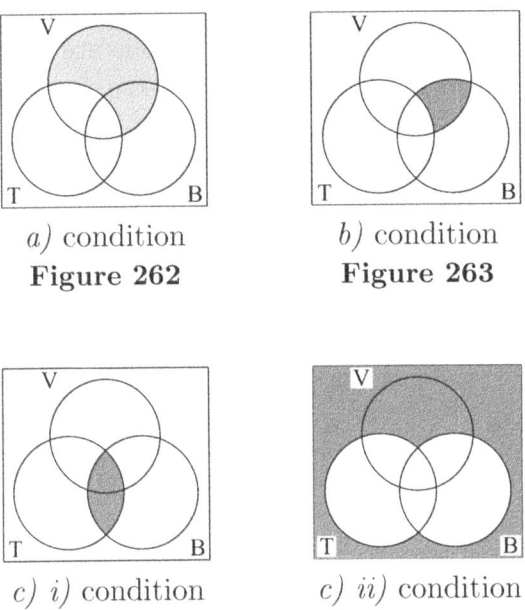

a) condition
Figure 262

b) condition
Figure 263

c) i) condition
Figure 264

c) ii) condition
Figure 265

By comparing the first two pictures, we can deduce that there must be a city with only tram as the means of transport. Consequently, the highlighted part in the fourth picture cannot be empty; therefore the highlighted part in the third picture has to be empty.

Let us gather all the information we have collected so far into a diagram (see Fig. 266).

B. Hujter et al., *Mathematical Explorations for Young Minds*, Problem Books in Mathematics, https://doi.org/10.1007/978-3-031-64185-5_27

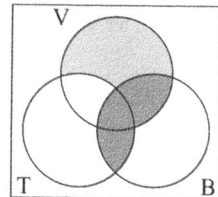

Figure 266

We can clearly deduce from the diagram there exists no city having both buses and trams.

12.2. The Octagon Shaped Rug

Solution. First divide the rug into parts as illustrated in Fig. 267.

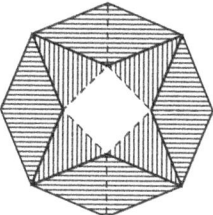 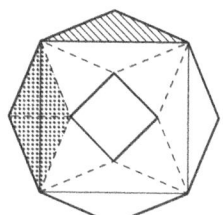

Figure 267   **Figure 268**

We claim that the 12 triangles with the striped pattern in Fig. 267 are all isosceles and congruent to each other.

The triangles with horizontal stripes share a side with the regular octagon, and each has another side that is a reflection of this one (based on the problem's conditions). Each of the triangles with vertical stripes has two sides that are also such reflections. Therefore, all 12 triangles are isosceles, and the equal sides of these triangles have the same length.

Two adjacent triangles with horizontal stripes can be combined together to form a rhombus, since all four sides are equal (we have marked one such rhombus using dots in Fig. 268). This rhombus shares an angle with the regular octagon, therefore it must be 135°, and consequently the remaining two angles are 45°. The missing angle between two such rhombuses is $135° - 2 \cdot 45° = 45°$, and thus the apex angle of the isosceles triangle between two rhombuses is also 45°. Thus, all 12 triangles are congruent.

Consider the triangle with skew stripes in the second picture. This is also half of the rhombus, obtained by dividing it across the longer diagonal. Therefore, the area of this triangle is the same as the areas of the 12 triangles in Fig. 267.

All these imply that if A denotes the area of a striped triangle, then the total area of the gray and black parts of Lidek's rug is $12A$. The combined area of the darker parts is $4A$; consequently, the remaining $8A$ is the area of the lighter parts. Thus, the area of the dark gray part equals half the area of the light gray part; therefore the answer is $\frac{1140}{2}$ cm² $= 570$ cm².

12.3. Filling in Tables with 1s and 2s

Solution. a) Using the two given digits, we can create four different numbers: 11, 12, 21, and 22. They give different remainders modulo 4, and therefore these exact four numbers have to appear in the table. However, it is not possible that one of 11 and 22 will appear in a row, and the other one in a column. Therefore, we can assume that both appear in a row. However, in this case the two columns turn out to be the same, and thus the four numbers we obtained are not even different.

b) Only two numbers can be obtained that are divisible by three: 111 and 222. These give different remainders modulo six, and therefore both should appear in the table. Similarly to part a), it is not possible that one of them appears in a row, while the other one in a column. Therefore, we can assume that both appear in a row.

Since there are only two types of digits, the third row will contain two equal digits; however, this implies that the numbers formed in the columns cannot be different. Therefore, it is not possible to fill out the table.

c) The modulo 8 remainder is determined by the last three digits of a number. Numbers 111, 112, 121, 122, 211, 212, 221, 222 all have different remainders modulo 8; therefore, all of them must feature somewhere in the table. With a little trial and error, this turns out to be possible, see Table 15.

1	2	1	1
2	1	1	1
2	2	1	2
2	2	2	1

Table 15

12.4. Pairwise Products

Solution. Let a, b, c, and d denote the four numbers. The six products can be divided into three pairs of equal products: $ab \times cd = ac \cdot bd = ad \times bc$.

Pairing up the products yields and multiplying them together yields
600, 800, 1000, 1200, 1200, 1500, 1800, 2000, 2400, 3000.

Only products $20 \cdot 60$ and $30 \cdot 40$ are equal. Thus the missing product must be the pair of 50, and its value can be easily calculated as $\frac{1200}{50} = 24$.

"I've found the original numbers: 4, 5, 6 and 10" said Tarkal. "I think there are no other solutions."

"Would you bet on it that there are no other possibilities?" asked Zsordi.

"Yes, I would bet a bar of chocolate on it," answered Tarkal.

Zsordi's Solution. We can assume that $a \leq b \leq c \leq d$. In Fig. 269, the pairwise products can be illustrated similarly to what we have seen in Problem 7.3.

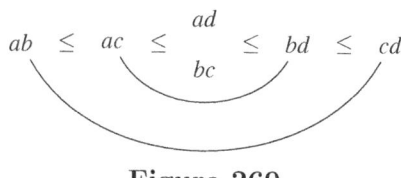

Figure 269

This way we get $ab = 20$, $ac = 24$, $bd = 50$, $cd = 60$. If we divide the first two products with each other, we obtain $\frac{ac}{ab} = \frac{c}{b} = \frac{24}{20} = \frac{6}{5}$.

We do not know whether ad or bc is the larger product, and therefore we have to consider two cases. *Case 1:* $bc = 30$. In this case, $c^2 = bc \cdot \frac{c}{b} = 30 \cdot \frac{6}{5} = 36$, so $c = 6$. Now we can find the original numbers: $b = 5$, $a = 4$, $d = 10$.

Case 2: $bc = 40$. In this case, $c^2 = bc \cdot \frac{c}{b} = 40 \cdot \frac{6}{5} = 48$, so $c = \sqrt{48} = 4 \cdot \sqrt{3}$. Now we can find the original numbers: $b = \frac{10}{3} \cdot \sqrt{3}$, $a = 2 \cdot \sqrt{3}$, $d = 5 \cdot \sqrt{3}$.

There is no other case, so the original four numbers can be either $4, 5, 6, 10$, or $2 \cdot \sqrt{3}, \frac{10}{3} \cdot \sqrt{3}, 4 \cdot \sqrt{3}, 5 \cdot \sqrt{3}$.

"I admit I wasn't right," said Tarkal. "What kind of chocolate should I get you?"

12.5. Numbers with an Alternating Sequence of Divisors

a) When the hikers attempted to solve this problem, they began by checking all numbers up to 20. They discovered that the following numbers have an alternating sequence of divisors:

2 (1, 2) **6** (1, 2, 3, 6) **10** (1, 2, 5, 10) **14** (1, 2, 7, 14) **18** (1, 2, 3, 6, 9, 18)

Tarkal immediately came up with a conjecture:

"I think that exactly those numbers have an alternating sequence of divisors that are even, but not divisible by 4."

"I wouldn't bet on it," Zsordi answered. She quickly found a counterexample: "The sequence of divisors of 30 (1, 2, 3, 5, 6, 10, 15, 30) is not alternating, so Tarkal's guess is not right."

"I also understand why 30 doesn't work," Albrecht continued. "When we multiplied 6 by 5, the original divisors (1, 2, 3, 6) were augmented with their five-folds: 5, 10, 15, 30. Since we've multiplied by an odd number, the new divisors also form an alternating sequence. However, when we've combined the two groups together, the property of being alternating got cancelled out (I highlighted the original divisors by underlining them)."

<u>1</u>, <u>2</u>, <u>3</u>, 5, <u>6</u>, 10, 15, 30

"However, if we multiply by 7 instead of 5," Tarkal's eyes lit up, "the first new divisor will come after the last original divisor, therefore, we do get an alternating sequence of divisors."

<u>1</u>, <u>2</u>, <u>3</u>, <u>6</u>, 7, 14, 21, 42

I already see a solution to part *a)*.

 Tarkal's Solution. $1806 = 42 \cdot 43$ has an alternating sequence of divisors with more than 10 divisors. Since 43 is a prime, the divisors of 1806 are the divisors of 42 (1, 2, 3, 6, 7, 14, 21, 42) and also their 43-folds. These also form an alternating sequence of divisors, and all these divisors are bigger than 42. Thus, the sequence of divisors of 1806 is

$$1, 2, 3, 6, 7, 14, 21, 42, 43, 2 \cdot 43, 3 \cdot 43, 6 \cdot 43, 7 \cdot 43, 14 \cdot 43, 21 \cdot 43, 42 \cdot 43,$$

or, after completing the multiplications,

$$1, 2, 3, 6, 7, 14, 21, 42, 43, 86, 129, 258, 301, 602, 903, 1806.$$

The resulting sequence of divisors is indeed alternating and comprises 16 divisors, which is more than 10. And, naturally, the solution still works if we keep prime factors 2, 3, and 7, but replace 43 with a larger prime factor.

Meanwhile, Zsordi found a solution with only two different prime factors.

 Zsordi's Solution. $486 = 2 \times 3^5$ is a solution to the problem, since its sequence of divisors is

$$1, 2, 3, 6, 9, 18, 27, 54, 81, 162, 243, 486.$$

In general, $n = 2 \times 3^k$, or in fact any number of the form $2 \cdot p^k$ (where p is an odd prime) has an alternating sequence of divisors. The divisors in increasing order are:

$$1, 2, p, 2p, p^2, 2p^2, p^3, \ldots, 2p^{k-1}, p^k, 2p^k.$$

Subsequently, the kids were trying to decide whether a number with an alternating sequence of at least 10 divisors smaller than 486 can be found. After a week, Albrecht found $342 = 2 \cdot 3^2 \cdot 19$, a number with an alternating sequence of 12 divisors:

$$1, 2, 3, 6, 9, 18, 19, 38, 57, 114, 171, 342.$$

The next day Tarkal found an even smaller solution:

$$1, 2, 3, 6, 7, 14, 21, 42, 49, 98, 147, 294.$$

By a thorough analysis of the cases (or by using a computer), it can be verified that 294 is the smallest solution.

b) The kids worked on the question about the square numbers separately, and they came up with three different solutions.

 Albrecht's Solution. If I could prove that a number with an alternating sequence of divisors is never divisible by 4, I would be done. After all, a number with alternating sequence of divisors (greater than 1) is obviously even (since all its divisors would otherwise be odd), and we know that even square numbers must be divisible by 4.

However, if we consider an even number (let us denote it by n), the ascending sequence of its divisors would end in $\ldots, \frac{n}{2}, n$. There are no divisors between these two, since the quotient would be between 1 and 2, so it would not be an integer.[44] If n is divisible by 4, both $\frac{n}{2}$ and n would be even; therefore, n's sequence of divisors cannot be alternating.

 Tarkal's Solution. I proved it slightly differently that a number divisible by 4 cannot have an alternating sequence of divisors.

If a number is divisible by 4, two even divisors can be assigned to every odd divisor q, namely $2 \times q$ and $4 \times q$.[45] Thus we have assigned two even divisors to every odd divisor (and different odd divisors have different even divisors assigned to them); therefore, there are at least twice as many even divisors as odd ones.

However, numbers with an alternating sequence of divisors cannot have more even divisors than odd divisors, because the ascending sequence of divisors starts with an odd number (1), and the odd and even divisors are alternating.

Zsordi found a solution that does not rely on divisibility by 4. Instead, we will take advantage of the fact that perfect squares have an odd number of divisors (as discussed in Problem 1.3).

 Zsordi's Solution. Let n be a number with an alternating sequence of divisors. Since the divisors of n are alternating, and the smallest divisor is odd (being 1), thus divisors at the odd positions in the sequence are all odd.

If $n > 1$ is a perfect square, it has an odd number of divisors, therefore the largest divisor, which is n itself, would be also odd. However, in this case all the divisors of n would be odd; therefore its sequence of divisors cannot be alternating.

12.6. Duel of Five Cards Against Five Cards

This game is really similar to "Rock-paper-scissors with a twist" (1.6): In that game players also take turns selecting cards from their opponent, albeit the cards feature symbols rather than numbers.

However, when attempting to decipher the intricate rules of the game, we discover that cards numbered 1, 2, and 3 mirror the dynamics of rock, paper, and scissors: 2 beats 1, 3 beats 2, and 1 beats 3. Card numbers 4 and 5 complete the picture by each beating two other cards. Each number beats the number that is smaller by 1 and larger by 2 (considering 1 as the number following 5).

After this discovery Albrecht reread his notes for the rock–paper–scissors game, inspecting *skew pairs*.

---

[44] It is worth recalling factor pairs discussed in Problem 1.3.

[45] It is worth recalling the table seen in the solution of Problem 8.5.

Albrecht's Strategy. The 10 numbers are divided into the skew pairs illustrated in Fig. 270. The main feature of these pairs is that the card in front of me beats its pair at my opponent.

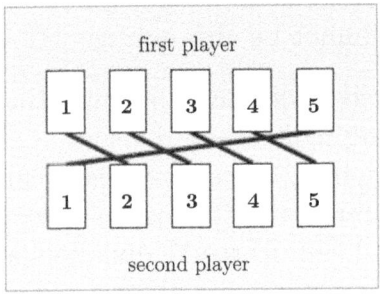

Figure 270

Playing as the second player, I always select the skew pair of the card that was picked by the first player. At the end of game a skew pair remains, and therefore I surely win.

"Ultimately you have to select the number that is one less than the number picked by your opponent," noted Zsordi.

"It's also possible to divide the $5+5$ cards into five *skew pairs* in different ways ensuring that the second player's card beats the first player's card," added Tarkal. "Each pairing will provide a different winning strategy for the second player."

It is worth telling the story behind this problem's creation. A mathematician from Oxisz wanted to create a more challenging variant of Problem 1.6.

He was familiar with the fact that the traditional rock–paper–scissors game can be expanded with two more symbols, see Fig. 271. The most renowned variation originates from an American blogger named Sam Kass,[46] and it gained global recognition through the TV show "The Big Bang Theory." In this version, the two new symbols are Lizard and Spock.[47]

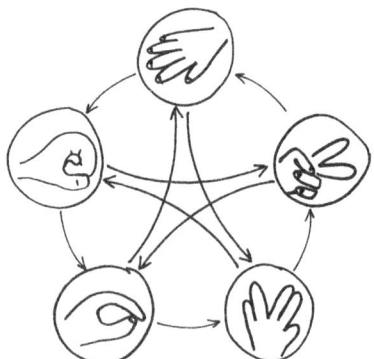

Figure 271

---

[46] The game is detailed on Sam Kass' website: http://www.samkass.com/theories/RPSSL.html.

[47] A character in the TV show "Star Trek." His customary greeting, known as the *Vulcan salute* (see Vulcan salute on Wikipedia), is accompanied by a characteristic hand gesture that sets it apart from the hand gestures associated with rock, paper, and scissors.

The rules are the following, as explained by Sheldon, a main character from "The Big Bang Theory": *"Scissors cuts paper covers rock crushes lizard poisons Spock smashes scissors decapitates lizard eats paper disproves Spock vaporizes rock crushes scissors."*[48]

If we assign numbers to the symbols in the following way:

$$1 = \text{rock} \quad 2 = \text{paper} \quad 3 = \text{scissors} \quad 4 = \text{Spock} \quad 5 = \text{lizard},$$

then we will get exactly the same dynamics as in our problem.

---

[48]It is also mentioned in the TV show that an advantage of this game is that draws occur less often than in the original game.

13. Bloody Serious Problems

13.1. Bidding on Each Other

Albrecht's Solution. Let us start by working backward. Dobda's last statement reveals that he scored at least one goal, and therefore Záhý scored at least $4 \times 1 + 1 = 5$ goals. Since Dobda knows more capitals than the number of goals Záhý scored, he knows at least six capitals. Záhý knows more than three times as many capitals, and therefore he knows at least $3 \cdot 6 + 1 = 19$ capitals.

"We are finished, since Hungary has exactly 19 counties," interjected Tarkal.

"This is true, but in fact we don't need this piece of information," said Albrecht.

Let us assume that Záhý knows at least 20 capitals. In this case, according to Dobda's first statement, he scored at least 21 points in the last year's Dürer competition, but this means that Záhý scored at least 3×21 points. However, this is impossible, since it was given that the maximum number of points was 60.

Hence, we have showed that Záhý knows exactly 19 capitals.

13.2. Queen, Rook, Bishop, Knight

Solution. A possible solution for each of the parts *a)* and *b)* is illustrated in Figs. 272 and 273.

Figure 272   Figure 273

c) Let us imagine that we have managed to arrange the pieces on the board. The knight has to capture two other pieces. These pieces cannot capture the knight, since the bishop, the rook, and the queen cannot move like a knight.

Therefore, the knight can be captured by only one other piece, and this implies that the pieces cannot be arranged on the board in the desired manner.

13.3. Polygon with Many Right Angles

It is easy to find a "nice" example with eight right angles, see Fig. 274.

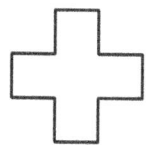

Figure 274

Is it possible to have more right angles in a 12-gon? Is it possible that this is the maximum?

Solution. Similarly to the solutions of Problems 4.4 and 8.3, we perform calculations with the angles. The sum of the internal angles of a 12-gon is $10 \cdot 180°$. If it contains k right angles, we can estimate the sum of the angles (using that the remaining angles are less than $360°$):

$$10 \cdot 180° < k \cdot 90° + (12 - k) \cdot 360° = 12 \cdot 2 \cdot 180° - k \cdot 270°.$$

Let us rearrange the inequality:

$$k \cdot 270° < 14 \cdot 180°.$$

Hence $k < \frac{28}{3} = 9\frac{1}{3}$, and therefore a 12-gon can have at most nine right angles.

Unfortunately, we have not yet solved the problem, since we have not decided whether a 12-gon can actually possess nine right angles. We need to find an example.

From the estimation we can derive that the average of the remaining angles has to be $330°$. This helps in finding an example.

We will show two 12-gons with an essentially different structure (naturally, other solutions are also possible), see Figs. 275 and 276.

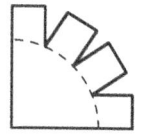 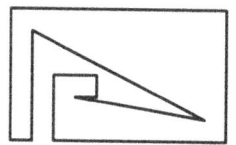

Figure 275   **Figure 276**

13.4. Lucky Numbers in Oxisz

"I'll tell you my discoveries," said Tarkal.

 Draft of Tarkal's Proof. First consider the odd numbers between 1001 and 2000, and these coincide with their largest odd divisors. Therefore, these numbers will be included in the sum:

$$1001, 1003, 1005, \ldots, 1999.$$

In the next step consider those numbers that are even, but not divisible by 4. If we divide these numbers by 2, we get their largest odd divisors:

$$501, 503, 505, \ldots, 999.$$

By considering numbers divisible by 4, but not divisible by 8, we get the following sequence of lucky numbers: $251, 253, 257, \ldots, 499$.

Let us collect the numbers we have found so far:

$$251, 253, \ldots, 499 \mid 501, 503, \ldots, 999 \mid 1001, 1003, \ldots, 1999.$$

Note that every odd number between 251 and 1999 occurred exactly once. We can be fairly certain that by continuing in this manner we will obtain each odd number between 1 and 1999 exactly once.

"Have you completed the calculation?" asked Zsordi.

"No, I got a bit confused with my numbers. But I'm fairly sure that we will get every number exactly once," answered Tarkal.

"I have a different approach that involves a lot less computation," said Albrecht.

 Albrecht's Solution. Let us focus on the condition of a given odd number appearing as the largest odd divisor.

First, let us find those numbers where 1 is the largest odd divisor. This is simple, since these are exactly the powers of 2:

$$1, 2, 4, 8, 16, 32, 64, 128, 256, 512, 1024, 2048, \ldots$$

Out of these only 1024 is between the given bounds.

Now let us find numbers with 3 being their largest odd divisors. These are exactly those numbers which can be written as 3 times a power of 2.

$$3, 6, 12, 24, 48, 96, 192, 284, 768, 1536, 3072, \ldots$$

Among these only 1536 is between the given bounds.

Now pick an arbitrary odd number between 1 and 1999, and denote it by $2k+1$. What can we say about numbers with $2k+1$ as their largest odd divisor? Clearly, they must be divisible by $2k+1$, so they can be written as $x \cdot (2k+1)$. On the other hand, x can only be a power of 2, since if it has an odd divisor different from 1, then the product would have a divisor larger than

$2k+1$. Thus we have concluded that exactly numbers of the form $2^n \cdot (2k+1)$ have $2k+1$ as their largest odd divisor.

$$(2k+1), 2 \cdot (2k+1), 2^2 \cdot (2k+1), 2^3 \cdot (2k+1), 2^4 \cdot (2k+1), \ldots$$

How many of these will be between 1001 and 2000?

The quotient of the consecutive members of this sequence is 2. Therefore, if a member is greater than 1000, the next member will be greater than 2000. Thus, at most one of these numbers can be between the given bounds.

On the other hand, if a member is less than 1001, the next one cannot be more than 2000; thus the sequence cannot jump over the given interval. This way we have proved that all these sequences contain exactly one number between the given bounds.

Consequently, we have proved that every odd number between 1 and 1999 will be the largest odd divisor of exactly one number.

Now we can quickly wrap up the solution by adding up odd numbers from 1 to 1999. This can be done with the method described in Problem 4.5, or by some other method. Ultimately, the sum of the lucky numbers is $1\,000\,000$.

"This solution reminds me Problem 8.5. We've also arranged divisors based on the powers of two there," suggested Zsordi.

"I haven't thought of that, but I concede that the similarity cannot be denied," admitted Albrecht.

13.5. Blood Donors and Raspberry Juices

a) We show that five glasses of raspberry juice can be achieved. Let us label the three people as A, B, and C. Blood donations are represented by arrows pointing from the donor to the recipient. They also indicate how the blood flows. At each step, we also record who has already received blood from whom, see from Figs. 277, 278, 279, and 280.

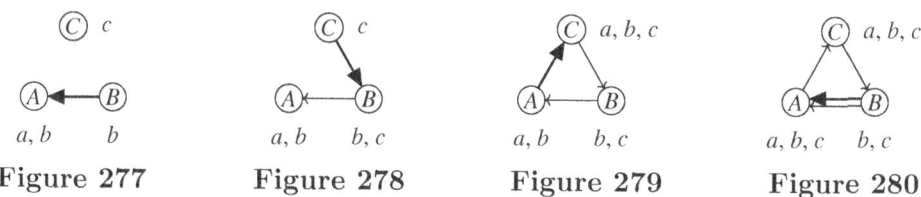

| Figure 277 | Figure 278 | Figure 279 | Figure 280 |

Since each person can receive blood from the other two, the maximum number of glasses they can achieve is 3×2. However, we prove that this maximum cannot be attained.

Let us focus on the last donation. The person who donated last cannot have all the blood; otherwise he could not have donated at all. Since this is the last donation, the donating person could not collect all the blood.

b) At most $6 \times 5 - 1 = 29$ glasses can be achieved. The argument is similar to the one in the previous part.

We will prove again that this theoretical maximum can be reached. We will generalize the construction from the previous part. To find the solution, we need to properly understand the dynamics with three people, as the donations will follow a similar pattern to those in the previous section.

The donations will be again of a cyclic structure. The orientation of the cycle is again opposite to the direction of the arrows.[49]

This time, there will be multiple cycles, divided into rounds. Each round will involve five donations. The first three rounds can be checked in Figs. 281, 282, and 283.

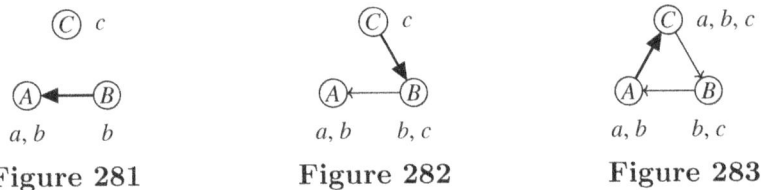

Figure 281   Figure 282   Figure 283

The blood donations will be carried out in a similar manner in the remaining two rounds.

Let us count the number of glasses. In the first round, the six people will obtain five glasses. In each subsequent round, the number of glasses increases by 6. Specifically, the 6th, 11th, 16th, and 21st donations yield two new glasses, while the remaining donations yield one. Therefore, in five rounds $5 + 4 \times 6 = 29$ glasses of raspberry juice will be obtained.

The solution above can be applied to any number, not just 6. n people can collect $n(n-1) - 1 = n^2 - n - 1$ glasses of juice in total.

The problem was indeed inspired by blood donation. Initially, we contemplated naming the donor and recipient *blood brothers*, but eventually opted for a different phrase. We have decided that the risk of misinterpretation was too high, as it suggested that the relationship was symmetric.

13.6. 37 Tokens in Three Heaps

"The game ends when each of the three heaps contains one token. In this case it is not possible to perform a legal move, while in all the other cases it is possible to make a legal move. Therefore, our aim is to create position $(1, 1, 1)$" Zsordi said in the beginning.

"We've already solved a simpler version of this game (7.6). In that game the strong positions were those where both heaps contained an even number of tokens. Let us try to find the strong positions in this game, too," continued Albrecht.

"It's probably worth starting with those positions where each heap contains an odd number of tokens," said Tarkal.

"If after our move each heap contains an odd number of tokens, then our opponent has to divide a heap with an odd number of tokens into a heap with

---

[49]This unexpected direction makes finding the construction difficult.

an odd and a heap with an even number of tokens. Thus after our opponent's round we'll receive two odd and an even heap. We can remove one of the odd heaps, and we can divide the even heap into two odd heaps. This way we can leave our opponent again with three odd heaps. Therefore during the game only we will create positions with three odd heaps, and thus ultimately we will be the one to create position $(1, 1, 1)$."

"Thus, the positions with three odd heaps are indeed strong positions. If we can reach such a position, we can win the game from there."

"And how can we guarantee reaching such a position first?"

"If there is both an even and an odd heap in the beginning, we have to go first, keep an odd heap, divide an even heap into two odd heaps, and remove the third heap."

If the initial position contains three odd heaps, we let the other player go first. And since there are 37 tokens in the beginning, it is not possible that each of the three heaps contains an even number of tokens."

"I want to remark that if we don't follow the above strategy, or could not decide whether to be the first player, a position with three even heaps can occur during the game. For example, it's possible that our opponent transformed position $(10, 12, 15)$ into position $(10, 8, 4)$ in his first move. What haven't yet established what to do in such a scenario."

"We will have some time to think about this on our trip home."

14. Icing on the Cake—Really Difficult Problems

14.1. Quadruples of Kritor, Lidek and Timol

Solution. We can assume that Kritor thought about numbers $a \leq b \leq c \leq d$, while Lidek thought about numbers $p \leq q \leq r \leq s$. Similar to Problem 7.3 we can arrange the pairwise sums in increasing order (except for the sums in the middle). Thus, the following have to be true:

$$a+b=p+q,\ a+c=p+r,\ b+d=q+s,\ c+d=r+s.$$

Let x denote the difference between the smallest members of the two quadruples, i.e., let $x = p - a$. Based on the first two equalities $q = b - x$ and $r = c - x$. From the last equality we can deduce $s = d + x$. Therefore, the second quadruple has to contain numbers $a + x, b - x, c - x, d + x$.

So far, we have analyzed four of the pairwise sums. We also know that the remaining two sums are the same; thus $a + d$ and $b + c$ have to be the same as $p + s$ and $q + r$ in a certain order.

Case 1: $a + d = p + s$ and $b + c = q + r$. Using what we have learned so far, we get $a + d = p + s = (a + x) + (d + x) = a + d + 2x$. Therefore, in this case $x = 0$, and thus the two quadruples have to be the same.

Case 2: $a + d = q + r$ and $b + c = p + s$. Substituting again yields $a + d = q + r = (b - x) + (c - x) = b + c - 2x$. Therefore, $x = \frac{(b+c)-(a+d)}{2}$. In this case, the second quadruple can be obtained as

$$\frac{a+b+c-d}{2}, \frac{a+b-c+d}{2}, \frac{a-b+c+d}{2}, \frac{-a+b+c+d}{2}.$$

In summary, either the quadruples of Kritor and Lidek have to be the same or Lidek's four numbers can be obtained uniquely from Kritor's numbers. However, the same can be said about Timol's quadruple. Therefore, we can find at most two kinds of quadruples for the six given sums, and this implies that they could not have considered three different quadruples.

Although we have answered the question of the problem, it is worth summarizing our discoveries. Let us introduce notation $S = \frac{a+b+c+d}{2}$ for quadruple (a, b, c, d). With this notation the pairwise sums of the following two quadruples are the same:

$$(a, b, c, d); \quad (S - d, S - c, S - b, S - a).$$

When will the two quadruples be different? A simple computation shows that these are different if and only if $a + d \neq b + c$. This means that if by taking the ascending orders of the six sums, the two sums in the middle are different, then these sums can be obtained in two different ways. However, if these two sums are the same, then they can be obtained in a unique way.

Using this observation, it is easy to generate further solutions to Problem 10.3. We only have to make sure that we have chosen quadruple $a < b < c < d$ with $a + d \neq b + c$.

14.2. Meandering Zigzag

If a meandering zigzag can be covered with n straight lines, then all of its inner breaking points (vertices) have to be the intersection of two of the straight lines.

We have already learned[50] that n straight lines can have at most $\binom{n}{2}$ points of intersection. The number of line segments in a meandering zigzag is one more than the number of breaking points. Thus if $\binom{n}{2} + 1 < 22$, then n lines are not enough to cover a meandering zigzag.

Since $\binom{6}{2} = 15$, six lines are not enough. However, $\binom{7}{2} = 21$, so we have a chance with seven lines. The seven lines have to form 21 distinct points of intersection, so no two lines can be parallel and no three can be concurrent. And the zigzag has to pass through all the 21 points of intersection, and it also has to change its direction at each.

"There's no guarantee that the the problem can be solved with just 7 lines," Tarkal remarked at this point. "Nonetheless, we could select 7 lines that form 21 points of intersection, and hope that we'll be fortunate enough to accommodate a meandering zigzag on them."

"And how should we choose the 7 lines? It would make me happy if our diagram would be clear," said Albrecht.

"Let's take the lines of the sides of a regular heptagon. This way we will get 21 points of intersection, and the picture will also be symmetric," suggested Zsordi.

---

[50] See the solution of Problem 8.2.

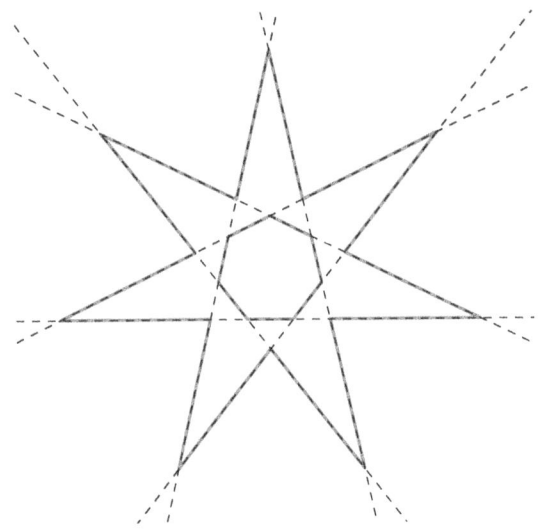

Figure 284

"Moreover, all the points of intersection are either on the heptagon, or on the outer polygon highlighted with a bold line (Fig. 284.) If we could join these two polygons with a single zigzag, all points of intersection would be covered," said Tarkal happily.

"This is not that hard!" exclaimed Zsordi, and she was proudly showing her drawing (Fig. 285): "Behold a meandering zigzag that can be covered with 7 lines."

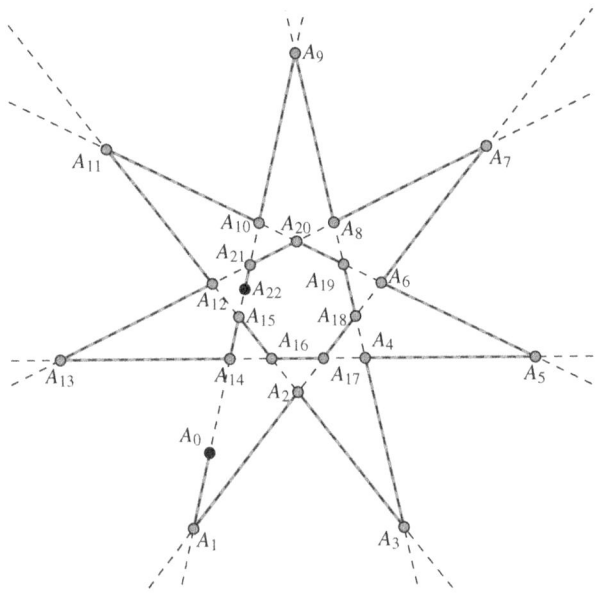

Figure 285

"The first and the last line segment of our zigzag are covered with the same line," noticed Albrecht. "Is this a coincidence? Or necessary? Could we also draw a zigzag that can be covered with 7 lines, and its first and last line segments are not on the same line?"

On their journey back from Oxisz, the team also addressed this question, and we hope to motivate our readers to do the same.

181

14.3. Mole and Bulldozer

The three friends recalled the simpler version of the problem (see Problem 6.4), where the mole-hills appeared in a 4 × 4 garden.

 Tarkal's Construction. We show an arrangement where each row and column contain exactly two mole-hills (Fig. 286).

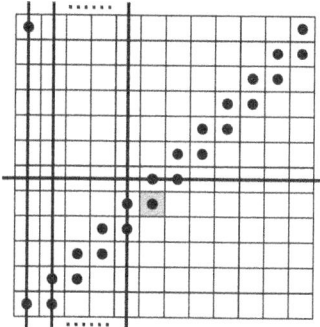

Figure 286

Why does this arrangement work? The farmer can bulldoze a total of twelve rows and columns, which is exactly the number of the mole-hills on the main diagonal. Therefore, the farmer can only decide for each mole-hill on the main diagonal whether he bulldozes it in a row or column.

Since the structure is symmetric, we can assume that the farmer chooses a column for the first mole-hill. There will be moment when a column is followed by a row (this moment can be seen in the picture above). However, in this case the farmer would not be able to bulldoze the mole-hill highlighted with gray under the main diagonal.

It is worth noting that our argument only used that the number of rows and columns to bulldoze is the same as the size of the garden, and at least one row and one column have to be bulldozed.

"The question remains whether the arrangement is *optimal*?" asked Zsordi.

"I'm almost sure that it's not optimal. If we would apply an *analogous* construction for a 4 × 4 garden, we would get a construction with eight mole-hills, and we've already seen that it's not optimal," replied Albrecht.

After giving it some thought, Zsordi came up with an idea.

 18 *Mole-Hills Are Not Sufficient.* We arrange the columns in a decreasing order based on the number of mole-hills they contain. Let us focus on the 7th column:

- If tit contains at least two mole-hills, then the first six columns contain at least 12 mole-hills. Therefore, the last six columns can contain at most six mole-hills.

- If it contains at most one mole-hill, then the last six columns contain at most six mole-hills.

182

Therefore, after bulldozing the six columns with the largest number of mole-hills, at most six mole-hills remain (in both cases), which can be bulldozed easily in the remaining six rounds.

"This also implies that if we can't bulldoze 19 mole-hills, then 7 rows and 7 columns have to contain 2 mole-hills each, and each of the remaining rows and columns have to contain 1 mole-hill,"—added Tarkal.

"Let's compare the construction with 7 mole-hills for the 4×4 garden, and the construction above. If we understand the difference between them, we will be able to improve the construction for the 12×12 garden," suggested Albrecht.

Albrecht's Solution. In Fig. 287, we show an example with $5 + 2 \times 7 = 19$ mole-hills.

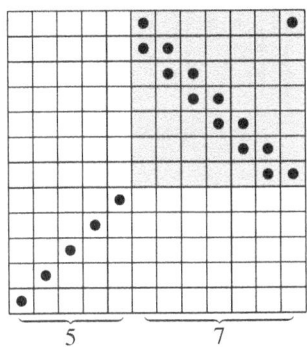

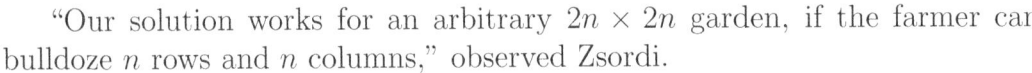

Figure 287

Why is this construction working? The farmer needs to use his bulldozer at least 5 times to bulldoze the mole-hills in the bottom left section.

Therefore, the farmer can use his bulldozer 7 times to take care of the section highlighted with gray. At the end of the previous part we have established that bulldozing such a structure requires at least eight passes, provided that there is at least one row and one column to bulldoze. And this is the case here, since there are seven rounds remaining, and the farmer can bulldoze exactly six rows and six columns.

This completes the solution of the problem, since we have provided a construction with 19 mole-hills, and proved that 18 mole-hills will not suffice.

"Our solution works for an arbitrary $2n \times 2n$ garden, if the farmer can bulldoze n rows and n columns," observed Zsordi.

"And what if we can bulldoze 4 rows and 8 columns in a 12×12 garden?" asked Tarkal.

"I think we'll be able to figure this out based on our previous ideas," Albrecht suggested.

Filling in the details is left to the reader.

14.4. Dividing the Cake

We can see that the pie is rectangle which has to be divided into pieces such that each has the same area and contains the same amount from the original perimeter.

The hikers found different kinds of solution to the problem.

Albrecht's Solution. a) The total length of the perimeter is $2 \times (18 + 36) = 108$ cm, and the third of which is 36 cm, which is precisely the length of the longer side. Let A and D denote the two vertices of a longer side, and let E denote the midpoint of the opposite side (see Fig. 288). These three points divide the perimeter into three parts of equal length. Our aim is to find point F inside the rectangle such that if we connect F to the three points, we obtain three equal areas.

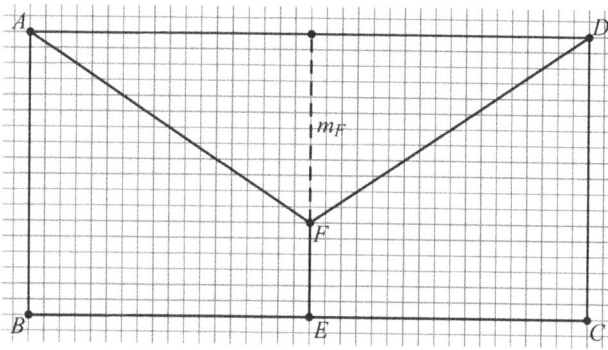

Figure 288

Selecting point F on the perpendicular bisector of the longer side ensures that the two parts on the side have equal area. Therefore, it is enough to ensure that the area of triangle AFD is equal to the third of the area of the rectangle, i.e., $18 \cdot 36/3 = 216$ cm². Since $AD = 36$ cm, we can determine the corresponding height, which must be 12 cm, or equivalently $EF = 6$ cm.

b) Let us divide each of the three (white) pieces from the previous part into two parts by bisecting both the area and the crust of the pie.

Dividing triangle AFD is easy. Let G be the midpoint of AD. Line segment FG is going to bisect both the area and the perimeter of the rectangle shared with the triangle, see Fig. 289.

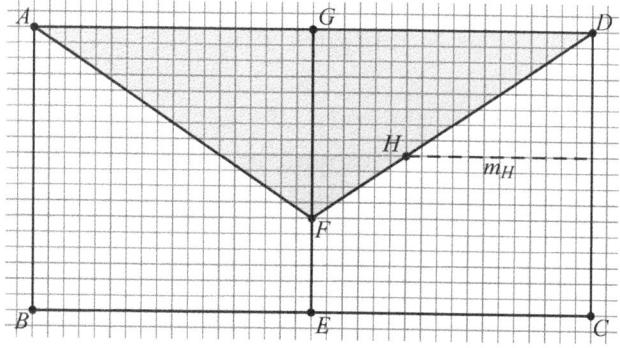

Figure 289

To bisect quadrilateral $ECFD$, we have to cut through C (to obtain 18 cm from the crust for both parts); therefore, we have to find point H on line segment DF such that line segment CH also bisects the area.

Once again, this can be done easily, since the area of triangle HDC is the sixth of the area of the rectangle, i.e., 108 cm^2. Since we know that the length of line segment CD is 18 cm, the corresponding height must be 12 cm. Further computations reveal that the distances of point H from sides AD and BC are 8 and 10 cm, respectively.

Bisecting quadrilateral $ABEF$ with line segment BI can be done in a similar way, see Fig. 290.

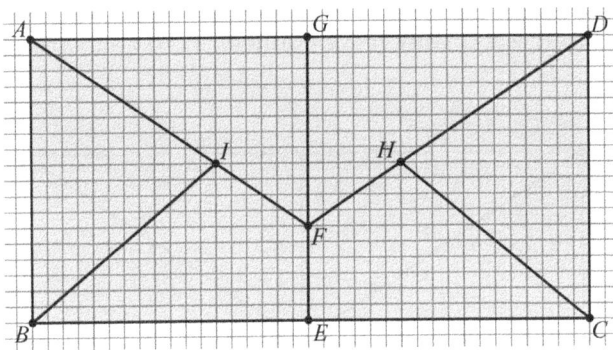

Figure 290

c) We will divide each piece from part *b)* in a way that will halve both the area and the crust. Triangles BIA, AFG, GFD, and DHC can be divided easily. Medians[51] IJ, FK, FL, and HM will bisect the area and the opposite side (i.e., the crust) simultaneously (see Fig. 291).

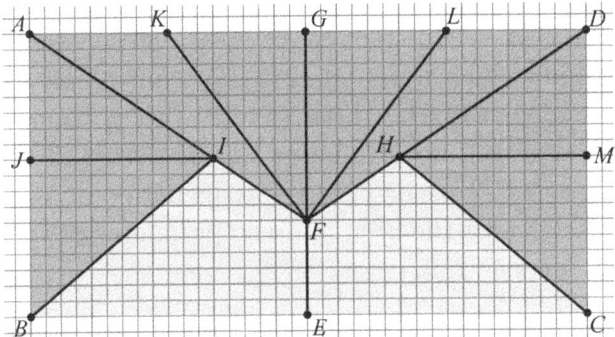

Figure 291

Now let us focus on quadrilateral $CHFE$ (see Fig. 292). If we manage to bisect it, we can do the same with quadrilateral $BEFI$. To bisect the shared part with the rectangle's perimeter (the crust), we have to bisect line segment CE. Let N be the midpoint of CE. We have to find point O such that NO bisects the area.

---

[51]The line segment connecting a vertex and the midpoint of the opposite side in a triangle is called a median.

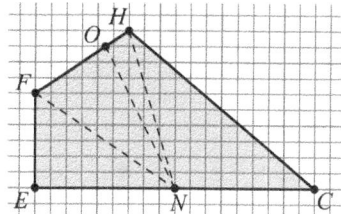

Figure 292

We can compute that the area of triangle NFE is 27 cm², while the area of triangle NCH is 45 cm². Both areas are smaller than the half of the area of quadrilateral $ECHF$, which is 54 cm²; therefore point O has to be on line segment FH (see figure).

We have to find O such that the area of triangle ONH is 9 cm² and the area of triangle OFN is 27 cm². The ratio of the two areas is 1 : 3; therefore O has to be chosen on side HF such that the ratio $HO : OF$ is also 1 : 3. This is because the two triangles share the altitude from vertex N; therefore, the ratio of their sides opposite N must be 1 : 3.

"I really like this process of halving," said Zsordi.

"It seems to me that this will work in general," added Tarkal.

"It will indeed," confirmed Albrecht. "If a convex polygon and point P on its perimeter is given, it's possible to find a line segment PB that bisects the area of the polygon."

"We only have to check that the method we've applied for quadrilateral $EHCF$ will work in general," established Zsordi. "First, let's connect point P to all the vertices of the polygon. This way we've divided our polygon into triangles. By determining the sum of their areas we can determine which triangle will contain point B, and then we can compute the ratio in which point B has to divide the given side. This method locates point B."

"That's right. Moreover, if the area has to be divided in a given ration instead of halving, our method will still work."

"I have a different kind of solution that does not rely on a halving process," said Tarkal. "I will divide it into 12 pieces first, and continue by creating bigger pieces from them."

 Tarkal's Solution. Choose point E such that its distances from sides AD, BC, and AB are all 6 cm. Similarly, let us choose point F such that its distances from sides AD, BC, and CD are all 6 cm. Furthermore, let us divide the perimeter into 12 equal parts and line segment EF into 6 equal parts (see Fig. 293).

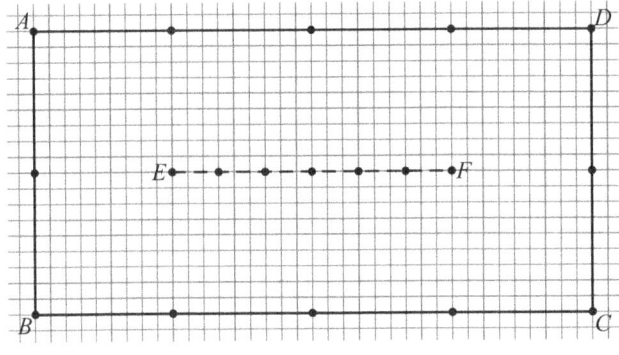

Figure 293

Now let us form some triangles, see Fig. 294.

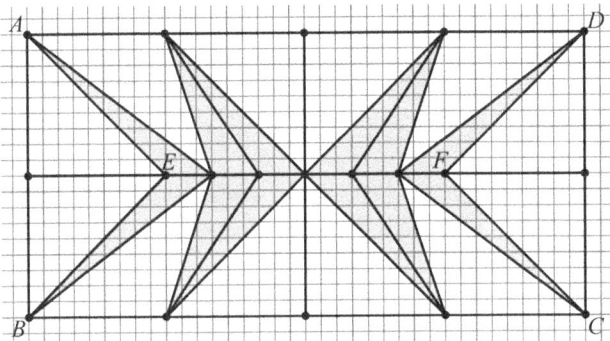

Figure 294

All the white triangles have a side of length of 9 cms on the perimeter of the cake. The heights corresponding to these sides are always 6 cms because of the choice of line segment EF. Therefore, all the white triangles have the same area.

Each of the gray triangles has a side with the same length, since line segment EF was divided into equal parts, and the corresponding height in all these triangles is 6 cms. Therefore, all the gray triangles have the same area, too.

If everybody gets a white and a gray piece, they get the same amount from the pie and also from the crust, as shown in Fig. 295.

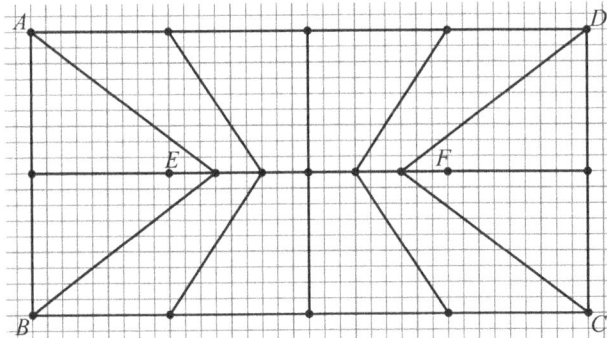

Figure 295

187

If everybody gets two gray and two white pieces or four gray and four white pieces, then we get a solution to parts *b)* and *a)*, respectively (see Figs. 296 and 297).

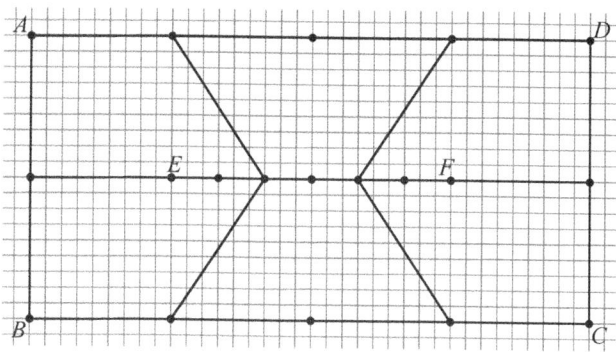

Figure 296

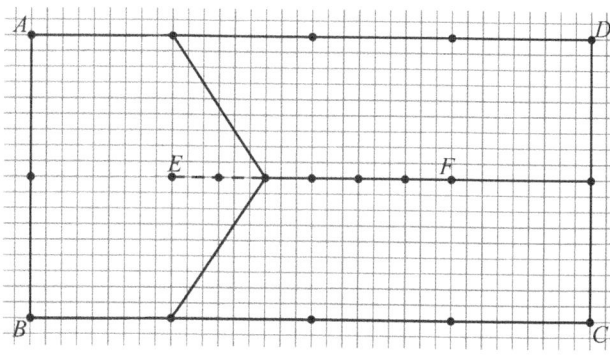

Figure 297

"You'll be surprised, that this method also has a connection to the halving process," said Albrecht.

"Does it?" wondered Tarkal. "And what is it?"

"You begin by dividing the cake into the 3 pieces in the last picture," began Albrecht. "Then, if you divide each piece with the method we discussed earlier, you obtain your solution with 6 pieces. If we bisect each piece again, we obtain your solution with 12 pieces."

"I admit, this is really surprising," said Tarkal.

"I also divided the cake into 12 pieces," said Zsordi. "I can create a solution with 6 pieces from it, however, my idea does not work for 3 pieces."

 Zsordi's Solution. Pick points E and F on diagonal BD such that they have the same distance from side CD and BC and also from side AB and AD (see Fig. 298). In other words, E must be on the angle bisector of angle BCD and F must be on the angle bisector of BAD.

Similarly to Tarkal's solution, let us divide the perimeter into 12 equal parts again. Let us connect the points above diagonal BD with point F, and the points below diagonal BD with point E.

188

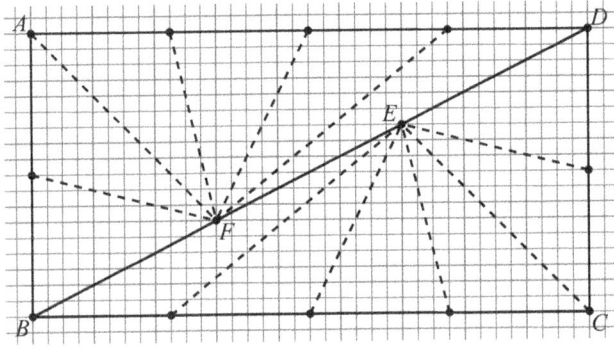

Figure 298

The sides of the resulting triangles on the perimeter of the rectangle have the same length (therefore ensuring the same amount of crust), and the corresponding heights will also be the same, because of the choice of points E and F. Therefore this division works.

If we only connect every second point, we obtain a solution to part *b)*, see Fig. 299.

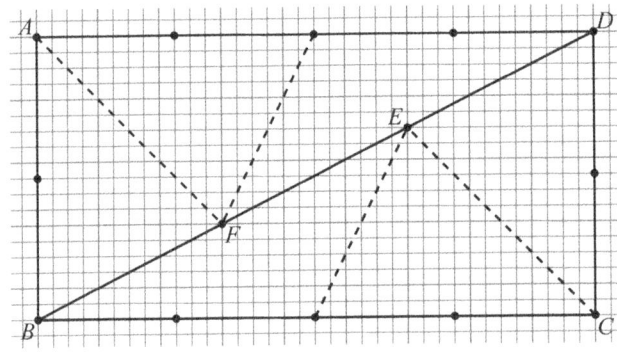

Figure 299

"Unfortunately, no matter how I tried putting together 3 pieces, I always obtained a concave piece," said Zsordi. "Therefore my solution is not as good as yours."

"Well, I think your solution is the best of our solutions," said Albrecht.

"How do you mean?" asked Zsordi.

"Let's generalize the problem," said Albrecht. "Let's say we want to divide the cake between n people instead of 3, 6 or 12. Furthermore, let the cake be an $a \times b$ rectangle. It's easy to see that your solution will work for all even values of n."

"Indeed, that's really nice," rejoiced Zsordi. "And what can be said when n is odd?"

"Maybe my method will work," wondered Tarkal.

The hikers managed to solve the problem on their way home by generalizing Tarkal's method. We encourage our readers to do the same.

14.5. Last Digits of Divisors

 Solution. Let us check the products of the last digits. Let $A_{1,9}$ denote the set of numbers ending in 1 or 9 and $B_{3,7}$ denote the set of numbers ending in 3 or 7. Let $T_{2,5}$ denote the set of numbers that are even or divisible by 5.

The last digit of the product of two numbers is determined by the last digits of the factors. For example, the product of a number from $A_{1,9}$ and from $B_{3,7}$ is always in $B_{3,7}$. This way we can create the following table of multiplication (Fig. 300):

	$A_{1,9}$	$B_{3,7}$	$T_{2,5}$
$A_{1,9}$	$A_{1,9}$	$B_{3,7}$	$T_{2,5}$
$B_{3,7}$	$B_{3,7}$	$A_{1,9}$	$T_{2,5}$
$T_{2,5}$	$T_{2,5}$	$T_{2,5}$	$T_{2,5}$

Figure 300

With the help of this table, in some cases it is possible to deduce the last digit of a factor from the last digit of the product. For example, the quotient of a number from $A_{1,9}$ and from $B_{3,7}$ always has to be in $B_{3,7}$.

First we will only focus on numbers with prime factors exclusively from $B_{3,7}$. Subsequently the remaining cases will be easier to handle.

According to Problem 10.5, every number has a prime sequence of divisors[52] starting with 1. Let us consider such a sequence and check the last digits of the divisors. In Fig. 301 a prime sequence of the divisors of $10647 = 3^2 \times 7 \times 13^2$ can be seen.

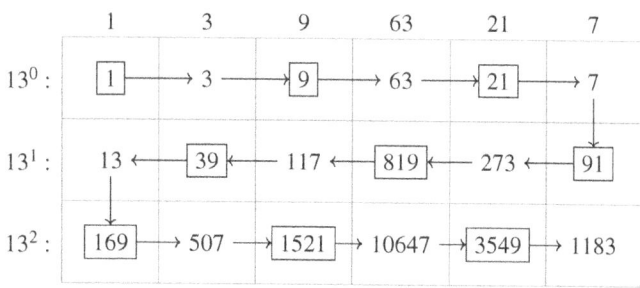

Figure 301

Note that every second divisor in the sequence is of type $A_{1,9}$. This is not specific to this case. In the general case we can also choose a prime sequence starting with 1. Every subsequent divisor is the result of multiplying or dividing by a prime of type $B_{3,7}$. As a result, the divisors of type $A_{1,9}$ and $B_{3,7}$ will be alternating in the sequence, as in Fig. 302.

---

[52] A sequence of divisors of a positive number is called a prime sequence if the quotient of the consecutive divisors is a prime.

$$A_{1,9} \longrightarrow B_{3,7} \longrightarrow A_{1,9} \longrightarrow B_{3,7} \longrightarrow A_{1,9} \longrightarrow B_{3,7} \longrightarrow \cdots$$

Figure 302

Since the sequence starts with a divisor of type $A_{1,9}$, the number of divisors of type $A_{1,9}$ cannot be smaller than the number of divisors of type $B_{3,7}$. Thus, we have proved the statement for numbers in which all prime factors are ending in 3 or 7.

Now let us consider an arbitrary positive integer $n = 2^s \times 5^t \times A \times B$, where A and B denote the product of the primes of types $A_{1,9}$ and $B_{3,7}$, respectively.

Note that numbers divisible by 2 or 5 cannot end in 1, 3, 7, or 9. Therefore, only those divisors of n can end in 1, 3, 7, or 9 that also divide $A \times B$. Let us consider the divisors of
$$n' = A \cdot B.$$

Every divisor of n' can be uniquely obtained as the product of a divisor of A and a divisor of B. Let the divisors of A be $a_1, a_2, \ldots, a_k$, while the divisors of B be $b_1, b_2, \ldots, b_\ell$. Let us group the divisors of n' based on their factors in A, see Table 16.

$$\begin{array}{cccccc}
a_1 b_1, & a_1 b_2, & a_1 b_3, & \ldots, & a_1 b_\ell \\
a_2 b_1, & a_2 b_2, & a_2 b_3, & \ldots, & a_2 b_\ell \\
\vdots & & & & \vdots \\
a_k b_1, & a_k b_2, & a_k b_3, & \ldots, & a_k b_\ell
\end{array}$$

Table 16

We have proved it earlier that at least half of the numbers $b_1, b_2, b_3, \ldots, b_\ell$ are of type $A_{1,9}$. Since multiplying by a_i does not change the type of a number, at least half of the numbers in group $a_i b_1, a_i b_2, a_i b_3, \ldots, a_i b_\ell$ are also of type $A_{1,9}$. This is true for all the groups separately, and therefore it is also true for their union, which is exactly the set of divisors of n'.

"I've found a slightly different solution," said Zsordi. "I would gladly share it with you."

"I'm eager to hear it," answered Tarkal.

"I've only considered prime factors of type $B_{3,7}$. First, let's try an actual number, e.g. $3969 = 3^4 \times 7^2$. Let's arrange its divisors into a table, as in Fig. 303.

	1	3	9	27	81
1	1	3	9	27	81
7	7	21	63	189	567
49	49	147	441	1323	3969

Figure 303

191

I have highlighted divisors ending in 3 or 7. This way we've obtained a chessboard pattern," said Zsordi.

"I see. So far this is the same as the original solution," interjected Albrecht.

"Yes, it begins the same way. We've seen that the divisors in the neighboring squares are in different groups, therefore their colours are different.

Thus we've created a chessboard pattern in the rectangle so that the top left corner is white. Is it true that the number of white squares cannot be smaller than the number of black squares?" asked Zsordi.

"If one side of the rectangle has even length, then it's easy to see that the number of black and white squares is the same," said Tarkal.

"Exactly," agreed Zsordi. "Our task is a bit more difficult if both sides have odd length. Let's visit all the squares in the order shown in Fig. 304.

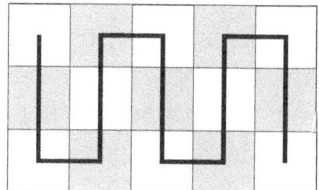

Figure 304

In this path the white and black squares are alternating. We have started from the top left corner, which is white. We have made an odd number of steps; therefore our journey will end at a white square. Thus we have proved there are one more white square."

"This line looks familiar, this is a prime order of divisors," shouted ALbrecht.

"Indeed," said Zsordi. "The situation changes a bit when we consider a number of the form $3^a 7^b 13^c$. Now we get a rectangular box with a chessboard pattern.[53] The question remains the same: is it true that the number of black fields cannot be larger than the number of white fields?"

"I like the way the problem was rephrased, and I'm happy to think about it," said Tarkal. "I find it very interesting that we do not have to talk about divisors at all."

"Feel free to think about it, this problem isn't especially difficult," answered Zsordi. "Moreover, the argument can be continued. When we consider numbers of the form $3^a 7^b 13^c 17^d$, we need to check the number of black and white fields in a four-dimensional rectangular box. I must admit, though, that I've struggled to properly visualize it."

14.6. Heap Dividing Game, an Arbitrary Number of Tokens in 3 Heaps

Albrecht and his team have already discovered in Problem 13.6 that positions with three odd heaps are strong positions.

---

[53]For example, considering number $3^2 7^2 13^2$ we get the cube and its coloring from Problem 8.6.

Positions with one or two odd heaps are weak positions, since a position with three odd heaps can be reached from them.

We still need to analyze positions with three even heaps. Our task is to identify the strong and weak positions among these.

On their way home from Oxisz the children started to work on this task.

They started by analyzing the positions with a small number of tokens. They identified the strong and weak positions among these (a quick reminder: A position is strong if only weak positions can be reached from it). They have identified the following strong positions with three even heaps: $(2, 2, 2)$, $(2, 2, 6)$, $(4, 4, 4)$, $(2, 2, 10)$, $(2, 6, 6)$, $(2, 2, 14)$, $(2, 6, 10)$, $(6, 6, 6)$, $(4, 4, 12)$, $(2, 2, 18)$, $(2, 6, 14)$, $(2, 10, 10)$, $(6, 6, 10)$, $(8, 8, 8)$, etc.

Looking for a pattern in these positions, Tarkal managed to formulate a conjecture about the strong positions in general:

(Tarkal's Conjecture) The strong positions are those triples (k, ℓ, m) where all three numbers have the same exponent for 2 in their prime factorization (i.e., the largest powers of 2 dividing them coincide).

Before arriving home, they have managed to prove the conjecture in two different ways. Albrecht initial observation was that reaching a position described by Tarkal is only possible by dividing a heap into two smaller heaps, where the exponents of 2 in their prime factorizations are identical. He termed these types of sums *balanced*.

He continued by checking the exponent of 2 in the addends of the balanced sums of 40:

- It can be divided into the sum of two odd numbers (the exponent is 0), e.g., $1 + 39$, $7 + 33$, or $17 + 23$.

- The exponent can be 1, e.g., $2 + 38$ or $10 + 30$.

- The exponent can be 2 in sums $4 + 16$ or $20 + 20$.

- However, 3 cannot be the exponent. There are only two possible addends that are smaller than 40: 8 and 24, but these cannot be combined into 40.

- Naturally, the exponent cannot be larger than 3, since otherwise both addends would be divisible by 16, while the sum is not divisible by 16.

Albrecht generalized his previous observations in a lemma he termed **The lemma of the balanced sums**.

Let the exponent of 2 in the prime factorization of n be a. Then:

(1) For an arbitrary $b < a$, n can be written as a balanced sum where the exponent of 2 in the prime factorization of the two addends is b.

(2) n cannot be written as a balanced sum where the exponent of 2 in the prime factorizations of the two addends is a.[54]

 The Lemma's Proof. (1) Since dividing n by 2^b the quotient is an even number, it can be divided (usually more than one way) into a sum of two odd numbers, $t_1 + t_2$. Then $2^b \cdot t_1 + 2^b \cdot t_2$ is a balanced division of n into two numbers.

---

[54]It is easy to see that the same holds for exponents $b > a$. However, this is not needed later; therefore, we have decided to exclude it from the lemma.

(2) Let us suppose a balanced division exists. Then the two addends can be written as $2^a \cdot t_1$ and $2^a \cdot t_2$, respectively, where t_1 and t_2 are odd. In this case
$$n = 2^a \cdot t_1 + 2^a \cdot t_2 = 2^a \cdot (t_1 + t_2);$$
however, $t = t_1 + t_2$ is even, and thus n has to be divisible by 2^{a+1}. Therefore, such a division cannot exist.

Albrecht proved the conjecture of Tarkal with the help of the lemma.

 Albrecht's Proof. We have to check three things:
- The only ending position of the game $(1,1,1)$ is indeed a strong position.

- Only weak positions can be reached from a strong position.
  In a strong position the size of the heaps can be written as $k = 2^a \cdot t$, $\ell = 2^a \cdot u$, and $m = 2^a \cdot w$, where t, u, w are odd numbers. In the untouched heap the exponent of 2 remains a. The second part of the lemma implies that the other heap cannot be divided into two parts where the exponent of 2 is a; therefore the resulting position cannot be strong.

- A strong position can be reached from every weak position.
  Let the sizes of the three heaps be $k = 2^a \cdot t$, $\ell = 2^b \cdot u$, and $m = 2^c \cdot w$ in a weak position, where t, u, w are odd numbers, and $a \geq b \geq c$. Since the position is weak, $a > c$ also holds.
  The first statement of the lemma implies that k can be written as a balanced sum where the exponent of 2 in the prime factorizations of both addends is c. If I divide the heap of size k into two parts in this way and remove the heap of size ℓ, I arrive at a strong position.

Zsordi later found another proof, the essence of which is the following.

 Draft of Zsordi's Proof. Observe that from a position with three even numbers it is only possible to move to another position with only even numbers; otherwise, we would end up with even and two odd heaps, and we already know that this is a weak position.

There is only a single position with only even numbers from which it is not possible to move to another position with only even heaps, namely $(2,2,2)$, and therefore this is a strong position. When two smart players play against each other, they will only choose positions with three even numbers, competing with each other to reach position $(2,2,2)$.

Note that a game starting from position $(2k, 2\ell, 2m)$ and only taking positions with three even numbers is basically the same as the original game starting from position (k, ℓ, m). Consequently, $(2k, 2\ell, 2m)$ will be a strong position (in both the original and the modified game) if and only if (k, ℓ, m) is a strong position in the original game.

From this observation, it is easy to deduce Tarkal's conjecture about the positions only containing even heaps.

Sources

In this chapter, we indicate the origin of each problem. If the author is known, we include their name. Despite the inevitable errors, we consider it important to preserve the name of each problem's author.

Additionally, we use codes to indicate in which Dürer Competition a problem appeared. The first number in the code indicates the competition season (the 1st season was in 2007–08, and the 14th season was in 2020–21). The following letter indicates the round: O—online, H—helyi (local), K—kifejtős (first round of finals, elaborative), V—váltó (second, relay round of finals). Next, the category letter is represented as follows: A—5–6th grade, B—7–8th grade, C and Cp—9–10th grade, D and Dp—11–12th grade, E—high school. Finally, the last number indicates the position of the problem within the problem set (strategy games are indicated by the letter J). For example, the code [10HA8] represents the 8th problem set in category A at the local round of the 10th Dürer Competition. The problem sets are available (in Hungarian) on the competition's website: https://durerinfo.hu/archivum/feladatsorok/.

The abbreviation KöMaL refers to the Középiskolai Matematikai és Fizikai Lapok—Mathematical and Physical Journal for Secondary Schools. KöMaL has been instrumental in organizing prestigious correspondence competitions for high school students, enhancing Hungarian high school education:

1.1. [10HA8] Proposed by Kartal Nagy.

1.2. [10HA4] Proposed by the authors.

1.3. [10HB7] Folklore.

1.4. [10VB14] Proposed by Kartal Nagy.

1.5. [10KB5] Proposed by the authors.

1.6. [7KAJ] This type of rock–paper–scissors was heard by one of the authors from a friend who thought it was equivalent to the original rock–paper–scissors.

2.1. [3VB3] Folklore.

2.2. [9VB5] Folklore.

2.3. [12HA7] Proposed by Kartal Nagy and authors.

2.4. [3KB5] Based on Problem B.3749. of KöMaL in September 2004. The story is written by the authors.

2.5. [13KA5] Proposed by the authors.

2.6. [11KBJ] Proposed by the authors.

3.1. [8HB4] The problem was inspired by an FIS Alpine Ski World Cup 2014 broadcast.

3.2. [7KB1] Folklore. In many places, however, there is no nice solution to the problem.

3.3. [11KB3] Proposed by Bálint Homonnay based on an English competition problem.

3.4. [11HA13] Proposed by the authors. It was inspired by Problem 18.11 in G. Blénessy, S. Dobos, T. Fazakas, A. Hraskó, Gy. Rubóczky (eds.), *Matkönyv: Kombinatorika 7-8.* (2005/2020), see matkonyv.fazekas.hu.

3.5. [12KA5] Proposed by the authors.

3.6. [13KAJ] Proposed by the authors.

4.1. The problem originates from the Hungarian novel *Strange Marriage* by Kálmán Mikszáth. It had also been Problem C.575. of KöMaL in March 2000.

4.2. [13KA2] Proposed by Luca Szepessy.

4.3. [12VA10] Proposed by the authors.

4.4. [10VC6] Proposed by Márton Dücső.

4.5. [6KB5] In this form, proposed by the authors.

4.6. [13KBJ] Proposed by the authors.

5.1. [10HB1] Proposed by Kartal Nagy.

5.2. [12HB8] Folklore. It went viral on the Internet in 2018 as "Homework in China for elementary school students."

5.3. [7KA3] Proposed by the authors. It was inspired by a question about alarm clocks, which was proposed by László Miklós Lovász. This original problem appeared around 2000 in the ABACUS International Math Competition (organized by Tivadar Divéki of Grace Church School, New York).

5.4. [13KB4] Proposed by the authors.

5.5. [6LA5] Folklore.

5.6. [6KAJ] Folklore.

6.1. [13OC6] Proposed by Kartal Nagy.

6.2. Problem 4 of the Lóránt Pálmay Math Competition in 2017.

6.3. [6LB3] Proposed by Máté Gyarmati.

6.4. [6KC2] Folklore.

6.5. Proposed by Kartal Nagy for Kalmár László Math Competition in 2020. The story of the problem has been altered.

6.6. [10KBJ] Proposed by the authors.

7.1. [12HC1] The problem is contained in the book *So You Think You've Got Problems?* by Alex Bellos.

7.2. Proposed by the authors.

7.4. [12HC3] Proposed by Lilla Tóthmérész.

7.3. Proposed by the authors.

7.5. [12VA14] Proposed by Kartal Nagy.

7.6. [8KAJ] Proposed by the authors. Games like this are known in the literature as take-and-break games.

8.1. [11VA9] Proposed by the authors.

8.2. Folklore.

8.3. [9VC7] Problem C.586. of KöMaL in May 2000.

8.4. [9KB4] Proposed by the authors.

8.5. [9HCp1] This problem was invented during the creation of the book.

8.6. [9KAJ] Proposed by the authors. The motivation came from a lecture on similar problems in graph theory, given by Gábor Kun at the IX. Dürer Final.

9.1. [11HC1] Proposed by Márton Dücső.

9.2. [11KD1] Proposed by the authors. The problem was born while thinking about another problem, the harder one was Problem A.764. of KöMaL in December 2019.

9.3. [5KB5] Folklore. The general question is also studied in mathematics research, see: L. Rédei and A. Rényi: on the representation of $1, 2, \ldots, N$ by differences. *Recueil Mathématique*, T. 61, 194.

9.4. Folklore.

9.5. [9KA5] Folklore.

9.6. [12KAJ] Proposed by the authors.

10.1. [12VA7] Folklore, we encountered the problem in a pub quiz.

10.2. [10KB2] Proposed by the authors. Part b) was problem B.5136. of KöMaL.

10.3. Proposed by the authors.

10.4. [7VB12] Folklore.

10.5. This problem was invented during the creation of the book.

10.6. [12KBJ] Proposed by the authors, based on the disappearing tic-tac-toe game.

11.1. [11HA10] Proposed by Kartal Nagy.

11.2. [8HB5] Folklore.

11.3. [13OD7] Proposed by the authors, the question was based on a real-life tile pattern.

11.4. [8HA15] The problem originates from a problem of Kvant, a Russian math journal for school students and teachers. The story has been altered. A generalized version was Problem A.780. of KöMaL.

11.5. [5LB4] Folklore.

11.6. [11KCJ] Folklore.

12.1. [4LB1] Proposed by unknown organizer.

12.2. [13KB5] Proposed by Kartal Nagy.

12.3. [12KC3] Proposed by Lilla Tóthmérész.

12.4. [10VD4] Folklore.

12.5. Proposed by the authors. Originally appeared at the Kalmár László Math Competition in 2020.

12.6. [7KBJ] Proposed by the authors.

13.1. [13HC1] Proposed by the authors.

13.2. [13KC2] Proposed by Kartal Nagy.

13.3. [12VC5] Proposed by the authors.

13.4. Proposed by Zoltán Gyenes. Originally appeared at the Kalmár László Math Competition in 2018.

13.5. [12HC5] Proposed by the authors and Dániel Tüzes.

13.6. [8KBJ] Proposed by the authors.

14.1. This problem was invented during the creation of the book.

14.2. [8VB15] Proposed by the authors.

14.3. Proposed by the authors. The problem can be considered as a special case of the Zarankiewicz problem.

14.4. [14HE2] Proposed by the authors. The answer is known for an arbitrary convex polygon.

14.5. [7KC5] Folklore.

14.6. [8KBJ] Proposed by the authors.

SPRINGER NATURE

GPSR Compliance

The European Union's (EU) General Product Safety Regulation (GPSR) is a set of rules that requires consumer products to be safe and our obligations to ensure this.

If you have any concerns about our products, you can contact us on ProductSafety@springernature.com

In case Publisher is established outside the EU, the EU authorized representative is:

Springer Nature Customer Service Center GmbH
Europaplatz 3
69115 Heidelberg, Germany

The manufacturer's authorised representative in the EU is Springer Nature Customer Service Centre GmbH, Europaplatz 3, 69115 Heidelberg, Germany. If you have any concerns regarding our products, please contact ProductSafety@springernature.com

Printed and bound by CPI Group (UK) Ltd, Croydon, CR0 4YY

27/03/2026

02079742-0001